190
Advances in Polymer Science

Advances in Polymer Science

Recently Published and Forthcoming Volumes

Surface-Initiated Polymerization II
Volume Editor: Jordan, R.
Vol. 198, 2006

Surface-Initiated Polymerization I
Volume Editor: Jordan, R.
Vol. 197, 2006

Conformation-Dependent Design of Sequences in Copolymers II
Volume Editor: Khokhlov, A. R.
Vol. 196, 2006

Conformation-Dependent Design of Sequences in Copolymers I
Volume Editor: Khokhlov, A. R.
Vol. 195, 2006

Enzyme-Catalyzed Synthesis of Polymers
Volume Editors: Kobayashi, S., Ritter, H., Kaplan, D.
Vol. 194, 2006

Polymer Therapeutics II
Polymers as Drugs, Conjugates and Gene Delivery Systems
Volume Editors: Satchi-Fainaro, R., Duncan, R.
Vol. 193, 2006

Polymer Therapeutics I
Polymers as Drugs, Conjugates and Gene Delivery Systems
Volume Editors: Satchi-Fainaro, R., Duncan, R.
Vol. 192, 2006

Interphases and Mesophases in Polymer Crystallization III
Volume Editor: Allegra, G.
Vol. 191, 2005

Block Copolymers II
Volume Editor: Abetz, V.
Vol. 190, 2005

Block Copolymers I
Volume Editor: Abetz, V.
Vol. 189, 2005

Intrinsic Molecular Mobility and Toughness of Polymers II
Volume Editor: Kausch, H.-H.
Vol. 188, 2005

Intrinsic Molecular Mobility and Toughness of Polymers I
Volume Editor: Kausch, H.-H.
Vol. 187, 2005

Polysaccharides I
Structure, Characterization and Use
Volume Editor: Heinze, T.
Vol. 186, 2005

Advanced Computer Simulation Approaches for Soft Matter Sciences II
Volume Editors: Holm, C., Kremer, K.
Vol. 185, 2005

Crosslinking in Materials Science
Vol. 184, 2005

Phase Behavior of Polymer Blends
Volume Editor: Freed, K.
Vol. 183, 2005

Polymer Analysis/Polymer Theory
Vol. 182, 2005

Interphases and Mesophases in Polymer Crystallization II
Volume Editor: Allegra, G.
Vol. 181, 2005

Interphases and Mesophases in Polymer Crystallization I
Volume Editor: Allegra, G.
Vol. 180, 2005

Block Copolymers II

Volume Editor: Volker Abetz

With contributions by
M. L. Arnal · V. Balsamo · C. Coenjarts · J.-F. Gohy
M. A. Hillmyer · M. Li · A. J. Müller · C. K. Ober

Springer

The series *Advances in Polymer Science* presents critical reviews of the present and future trends in polymer and biopolymer science including chemistry, physical chemistry, physics and material science. It is adressed to all scientists at universities and in industry who wish to keep abreast of advances in the topics covered.
As a rule, contributions are specially commissioned. The editors and publishers will, however, always be pleased to receive suggestions and supplementary information. Papers are accepted for *Advances in Polymer Science* in English.
In references *Advances in Polymer Science* is abbreviated *Adv Polym Sci* and is cited as a journal.

Springer WWW home page: http://www.springeronline.com
Visit the APS content at http://www.springerlink.com/

Library of Congress Control Number: 2005933605

ISSN 0065-3195

ISBN 978-3-642-06595-8 ISBN 978-3-540-31603-9 (eBook)
DOI 10.1007/978-3-540-31603-9

Springer is a part of Springer Science+Business Media

springeronline.com

Cover design: *Design & Production* GmbH, Heidelberg
Typesetting and Production: LE-TeX Jelonek, Schmidt & Vöckler GbR, Leipzig

Volume Editor

Prof. Dr. Volker Abetz

GKSS – Forschungszentrum Geesthacht GmbH
Institut für Polymerforschung
Max-Planck-Straße
21502 Geesthacht, Germany
Volker.Abetz@gkss.de

Editorial Board

Advances in Polymer Science Also Available Electronically

For all customers who have a standing order to Advances in Polymer Science, we offer the electronic version via SpringerLink free of charge. Please contact your librarian who can receive a password or free access to the full articles by registering at:

springerlink.com

If you do not have a subscription, you can still view the tables of contents of the volumes and the abstract of each article by going to the SpringerLink Homepage, clicking on "Browse by Online Libraries", then "Chemical Sciences", and finally choose Advances in Polymer Science.

You will find information about the

- Editorial Board
- Aims and Scope
- Instructions for Authors
- Sample Contribution

at springeronline.com using the search function.

Preface

Self-assembly of matter into ordered structures is an important issue in biology and materials science. For example, the intramolecular and intermolecular self-assembly of proteins is essential for their functions as enzymes or carrier systems. In materials science, the organization of matter is important for mechanical, optical, electrical, and other properties not only in bulk or thin film systems, but also in solution. Self-assembly is a property of some materials which can be used for the controlled generation of regularly structured materials. This self-assembly can occur on different length scales. Atoms, small molecules or repeating units of some polymers can aggregate into crystals, which have typical periodicities on the sub-nanometer range. On a much larger length scale another class of materials becomes interesting for its capability to self-assemble into ordered structures: block copolymers. They are a fascinating class of condensed soft matter. By linking different, mostly immiscible chains together chemically, they self-assemble often into crystal-like structures (supercrystals, microphase morphologies). The periodic length scale in microphase separated block copolymers is typically in the range between 10 and several 100 nm, i.e. 100 to 1000 times larger than the periodic length of atomic crystals.

Block copolymers have already attracted significant scientific and economic interest during the last few decades. A prerequisite for the formation of rather periodic microphase morphologies is a fairly narrow polydispersity both of composition and molecular weight. After the discovery of living anionic polymerization in the middle of the last century, the controlled synthesis of block copolymers became possible for the first time. Since then, other synthetic schemes with a large degree of control were developed, and this trend is still continuing. For this reason more and more monomers become eligible for incorporation into block copolymers, thus adding more possible functions into microphase-structured materials.

Although many books and reviews have been written about block copolymers, the ongoing research motivated us to give an overview of the results during the last years in various aspects of the field.

The topic is split into two volumes and is organized as follows. In the first chapter the progress in different synthetic routes to controlled block copolymers of various architectures will be presented, the second chapter tries to give an overview of the phase behavior of block copolymers in the bulk state and in concentrated solution. The interplay between crystallization on a segmental scale and the microphase separated structure of block copolymers with crystallizable blocks will be discussed extensively in the third chapter, since this aspect has not been addressed in such detail before in a review. After having treated bulk or bulk-like (highly concentrated solutions) states, the fourth chapter presents the structure formation of block copolymers in more dilute solutions, where various micellar superstructures can be found. The last two chapters deal with applications of block copolymer structures as precursors for the formation of mesoscale porosity (fifth chapter), and as precursors for the formation of controlled patterns on surfaces enabling new ways to lithography on a length scale, which is not accessible by other lithographic techniques thus far (sixth, last chapter).

We are aware that these two volumes do not cover all aspects of the research done on block copolymers, but nevertheless it is our hope that it will find interested readers and be a basis and stimulation for future further research on these materials.

Geesthacht, October 2005 *Volker Abetz*

Contents

Contents of Volume 189

Block Copolymers I

Volume Editor: Volker Abetz
ISBN: 3-540-26580-5

Adv Polym Sci (2005) 190: 1–63
DOI 10.1007/12_001

Published online: 5 October 2005

Nucleation and Crystallization in Diblock and Triblock Copolymers

Alejandro J. Müller (✉) · Vittoria Balsamo · María Luisa Arnal

Grupo de Polímeros USB, Departamento de Ciencia de los Materiales, Universidad Simón Bolívar, Apartado 89000, 1080-A Caracas, Venezuela
amuller@usb.ve

Dedicated to Prof. Estrella Laredo on the occasion of her 65th birthday.

Abstract Crystallization of block copolymer microdomains can have a tremendous influence on the morphology, properties and applications of these materials. In this review, particular emphasis is placed on the nucleation, crystallization, thermal properties and morphology of diblock and triblock copolymers with one or two crystallizable components. The issues of the different types of nucleation processes (i.e., homogeneous nucleation and heterogeneous nucleation by different types of heterogeneities and surface nucleation) and their relation to the crystallization kinetics of the components is addressed in detail in a wide range of polymeric materials for droplet dispersions, blends and block copolymers. The case of AB double crystalline diblock copolymers is discussed in the light of recent works on biodegradable systems, while the nucleation, crystallization and morphology of more complex materials like ABC triblock copolymers with one or two crystallizable components are thoroughly reviewed.

Keywords ABC triblock copolymers · Block Copolymers · Crystallization · Homogeneous nucleation

Abbreviations

AFM	Atomic force microscopy
aPP	Atactic polypropylene
DSC	Differential scanning calorimetry
HDPE	High-density polyethylene
iPP	Isotactic polypropylene
LLDPE	Linear low-density polyethylene
MD	Microdomain
ODT	Order–disorder transition
PB	Poly(butadiene)
PBO	Poly(oxybutylene)
PCL	Poly(ε-caprolactone)
PE	Polyethylene
PEO	Poly(ethylene oxide)
PEP	Poly(ethylene-*co*-propylene)
PI	Poly(isoprene)
PLLA	Poly(L-lactide)
POM	Polarized optical microscopy
PPDX	Poly(*p*-dioxanone)
PS	Polystyrene
PVCH	Poly(vinylcyclohexane)
SAXS	Small-angle X-ray scattering
SEB	Styrene-*ran*-ethylene-*ran*-butene
TEM	Transmission electron microscopy
TSDC	Thermally stimulated depolarization current
UB	Unmixed blend
WAXS	Wide-angle X-ray scattering
D_w^x, C_y^z	notation employed for block copolymers, where the subscripts (w and y) denote the composition in weight percent and the superscripts (x and z) the number-average molecular mass in kilograms per mole. In the case of PPDX-*b*-PCL diblock copolymers, the letter D is used for PPDX and the letter C for PCL

1 Introduction

The ability of block copolymers to self-assemble into organized microdomain (MD) structures when the thermodynamic repulsion between the constituents is high enough seems to be fairly well understood. This is particularly true in the case of amorphous diblock copolymers where phase diagrams for particular systems have been successfully predicted and experimentally proven [1–5].

Crystallization within block copolymer MDs is an important issue since it can completely change the block copolymer morphology. The structure development in semicrystalline block copolymers depends on two competing self-organizing mechanisms: microphase separation and crystallization. The most commonly studied of the semicrystalline block copolymer systems in the literature are AB diblock copolymers or ABA triblock copolymers, where one block is amorphous and the other semicrystalline. It is generally accepted that the changes of state as a function of temperature can determine the final morphology according to three key transition temperatures: the order–disorder transition (ODT) temperature, T_{ODT}, the crystallization temperature, T_c, of the crystallizable block, and the glass-transition temperature, T_g, of the amorphous block.

Five general cases have been described in the literature for AB diblocks with one crystallizable block:

1. *Homogeneous melt*, $T_{\mathrm{ODT}} < T_c > T_g$. In diblock copolymers exhibiting homogeneous melts, microphase separation is driven by crystallization if T_g of the amorphous block is lower than T_c of the crystallizable block. This generally results in a lamellar morphology where crystalline lamellae are sandwiched by the amorphous block layers and spherulite formation can be observed depending on the composition [6–10].
2. *Weakly segregated systems*, $T_{\mathrm{ODT}} > T_c > T_g$ *with soft confinement*. In this case, crystallization often occurs with little morphological constraint, enabling a "breakout" from the ordered melt MD structure and the crystallization overwrites any previous melt structure, usually forming lamellar structures and, in many cases, spherulites depending on the composition [10–18].
3. *Weakly segregated systems*, $T_{\mathrm{ODT}} > T_c < T_g$ *with hard confinement*. In this case, the crystallization of the semicrystalline block can overwhelm the microphase segregation of the MD structures even though the amorphous block is glassy at the crystallization temperature, because of the weak segregation strength [19].
4. *Strongly segregated systems*, $T_{\mathrm{ODT}} > T_c > T_g$ *with soft confinement*. If the segregation strength is sufficiently strong, the crystallization can be confined within spherical, cylindrical or lamellar MDs in strongly segregated systems with a rubbery block [10, 15–17, 20–28].
5. *Strongly segregated systems*, $T_{\mathrm{ODT}} > T_c < T_g$ *with hard confinement*. A strictly confined crystallization within MDs has been observed for strongly segregated diblock copolymers with a glassy amorphous block [29–42].

Examples of all these cases can be found in Table 1, where a thorough listing of most references dealing with block copolymers crystallization in the past decade is presented along with salient features of the systems that have been studied.

Since excellent reviews on block copolymer crystallization have been published recently [43,44], we have concentrated in this paper on aspects that have not been previously considered in these references. In particular, previous reviews have focused mostly on AB diblock copolymers with one crystallizable block, and particular emphasis has been placed in the phase behavior, crystal structure, morphology and chain orientation within MD structures. In this review, we will concentrate on aspects such as thermal properties and their relationship to the block copolymer morphology. Furthermore, the nucleation, crystallization and morphology of more complex materials like double-crystalline AB diblock copolymers and ABC triblock copolymers with one or two crystallizable blocks will be considered in detail.

In contrast to binary block copolymers, where one independent composition variable, ϕ_A, and one interaction parameter, χ_{AB}, are the parameters that determine the equilibrium phase morphology, the morphology of ABC triblock copolymers is governed not only by two independent composition variables (ϕ_A, ϕ_B, $\phi_C = 1 - \phi_A - \phi_B$), but also by the balance of three interaction parameters (χ_{AB}, χ_{AC}, χ_{CB}) which may be expressed alternatively by their interfacial tensions (γ_{AB}, γ_{AC}, γ_{CB}). As a consequence, chain topology (ABC, BAC, ACB) is of key importance. Owing to the combination of all these parameters, ABC triblock copolymers offer the opportunity to tailor fascinating new morphologies [45–50]. The formation of these mesophases has been theoretically analyzed by several authors. Mogi et al. [45] and Nakazawa and Otha [51] were able to explain some of the simplest structures. Thereafter, Kane et al. [52] reported a more elaborate extension of the Semenov theory to lamellar ABC triblock copolymers. Later, Zheng and Wang [53] calculated phase diagrams, which were successful in explaining some more complex morphologies. The morphologies of amorphous symmetric and asymmetric ABC triblock copolymers in the strong segregation limit were then well described by a simple thermodynamic model based on a Flory–Alexander–deGennes–Semenov approach. Using this approach, the experimentally discovered morphologies of linear polystyrene-*b*-poly(butadiene)-*b*-poly(methyl methacrylate), PS-*b*-PB-*b*-poly(methyl methacrylate), were explained in terms of a minimization of interfacial energy [54]. Matsen [55] used self-consistent field theory for the prediction of the morphology in ABC triblock copolymers. There are also reports that have predicted T_{ODT} in ABC triblock copolymers for different compositions and incompatibilities. These theoretical calculations have been based on the "mean-field" approach of Leibler for the weak segregation limit [56].

Although there are very comprehensive studies that explain the behavior of amorphous ABC triblock copolymers, this is not the case when one or more of the components are able to crystallize. In this case, a much more complex behavior is expected because of the interplay of crystallization–microphase separation.

Table 1 Crystallizable diblock and triblock copolymer data reported in recent literature

System [a]	M_n (kg mol^{-1})	Melt morphology [c]	Morphology [c]	Method [d]	Ref.
$EO_{76}MMA_{24}$	83 [b]	–	Lam	D,E	9
$E_{24\to75}EP_{76\to25}(4)$	80–180	Lam/Cyl/Sph	Lam	A,B,D	8
$E_{25\to75}EE_{25\to75}(4)$	20–45	Lam/Cyl/Sph	Lam		
$E_{26\to27}MB_{73\to74}(2)$	35 [b]–45	Dis/Cyl	Lam/Cyl	A,B,D	10 15 25
$E_{27}MB_{73}(2)$	63–88	Cyl	Cyl		
$E_{27\to70}VCH_{73\to30}(4)$	17–23	Cyl/Gyr/Lam	Cyl/Gyr/Lam	A,B	41
$VCH_{43\to17}E_{14\to67}VCH_{43\to17}(7)$	28–75	Sph/Cyl(In)/ Lam	Sph/Cyl(In)/ Lam	D,C,G	
$E_{28}VCH_{72}$	36	Cyl	Cyl	A,B,D	25
$E_{51}hhP_{49}$	50 [b]	Lam	Lam	A,B,D	13
$E_{46}S_{54}$	130	Lam	Lam	A,B	137
$S_{13}B_{74}S_{13}$	133	Cyl	Cyl	A,C	138
$C_{26\to61}B_{74\to39}(7)$	10–29	Lam	Lam	A,D,E	139 140
$C_{61}B_{39}/C^6$	11	Lam	Lam	A	139
$C_{47}B_{53}/C^7$	10	Lam	Lam		
C_9ET_{91}	–	–	Lam		141
$E_{15\to95}B_{85\to5}(13)$	40–281	–	–	C,D	142
$E_{50\to75}S_{50\to25}(6)$	120–240	–	–		
$E_{52}VCH_{48}$	15	Lam	Lam	A–C	12 143
$EO_{20}C_{80}$	17	–	Lam	A,B,D	144 145
$EO_{32\to72}BO_{68\to28}(25)$	2.5–3	Cyl/Gyr/ Lam/Dis/OS	Lam	A,B,D	18 146
$EO_{24,33}S_{34,53}EO_{24,33}(2)$	70–112	OS	Lam/Stes	A,B D–F	31 147
$EO_{75}C_{25}$	20	Hom	Stes	A,B, D,E G,H	135
$S_{4\to17}EO_{18\to74}C_{7\to78}(4)$	27–112	Hom	Stes/Ax/ PCcrys		
$EO_{69}BO_{31}$	2	–	Lam	A,H	148
$EO_{33}BO_{35}EO_{33}$	5	–	Lam		
$EO_{66}S_{34}$	16	Dis	Lam	A,B D,K	35
$S_{66\to87}C_{34\to13}/S^{0\to0.5-1}(3)$	12–23	Sph	–	A,D	149
$S_{66\to87}C_{34\to13}/Ch\ (3)$	12–23	Sph	Lam		

Table 1 (continued)

System [a]	M_n (kg mol^{-1})	Melt morphology [c]	Morphology [c]	Method [d]	Ref.
$S_{9\to35}B_{37\to6}C_{36\to77}(6)$	132–219	–	Lam/Co-Cont/ Cyl/C-S Cyl/ UL/Stes		97 98 115
$S_{9\to57}E_{6\to37}C_{16\to77}(6)$	137–219	–	Stes	C,D	116
$B_{20}C_{80}$	68	–	Stes		129 150
EOB	7 [b]	–	Lam	E,H	151
$EO_{20}C_{80}$	26	Sph	Lam		
$EO_{22}(C_{39})_2$	23	Sph	Lam	A,D	105
$C_{38}EO_{18}C_{38}$	22	Sph	Lam		
$C_{10\to61}B_{90\to39}(5)$	11–14	Dis/Lam	Lam	A,D	152–154
$C_{61}B_{39}/C^{8,22}$	12	Lam	Lam	A,D	152
$C_{47}B_{53}/C^{8}$	11	–	Lam		
$C_{26}B_{74}(2)$	18	Cyl	Lam/Cyl	A,B,D	155
$S_{74\to87}C_{26\to13}(2)$	12–23	–	–	A,D	156
$S_{27\to66}C_{73\to34}(3)$	12–30	Lam	Lam	A,C,D	156 157
$S_{70}C_{30}$	39	Sph	Sph	A,D,C	157
$EO_{45\to59}B_{41\to55}(4)$	7–9	Lam	Lam	D,E G,H	158
$E_{18\to28}VCH_{72\to82}(2)$	26–36	Sph/Cyl	Sph/Cyl	A,C D,G	159
$EO^{9}S^{9}$	18	Lam	Lam	A,B,L	160
$E_{18}EP_{64}E_{18}$	55	Hom	Lam	A–D	161
$BO_{12\to38}EO_{24\to76}BO_{12\to38}(19)$	4–12	Dis/Lam/ Cyl/Sph	–	A,G	162
$S_{15}EP_{70}E_{15}$	103	Cyl	Cyl	A–D	163
$C_{10\to19}B_{81\to89}(6)$	8–62	Cyl or Sph	Lam/Cyl or Sph	A,C,D	16
$E_{39}S_{61}$	28	HPL	–	A,C,G	164
$EO^{4}B^{21}$	25	Sph	Sph	D,H	23
$EO_{47}S_{53}$	18	Lam	Lam	A,C B,D,G	36 165
$EO_{47}S_{53}/S^{0.5\to5}(2)$	18	Cyl	Cyl	A,C B,D	37
$EO_{50}{}^{6}B_{50}{}^{5}/B^{0\to1}(5)$	11	Lam/Cyl/Sph	Lam	A,D,C	92
$THF_{29\to59}S_{41\to71}(4)$	30–37	Lam/Cyl/Sph	Stes	D,E	166

Table 1 (continued)

System [a]	M_n (kg mol^{-1})	Melt morphology [c]	Morphology [c]	Method [d]	Ref.
EO^5EI_{cy}	–	–	Lam		
$EO^5EI_{cy}DC$	–	–	Dis		
EO^5EI_{li}	–	–	Lam	A–D,M	167
$EO^5EI_{li}DC$	–	–	Lam		
EO^5EI_{br}	–	–	Lam		
$EO^5EI_{br}DC$	–	–	Lam		
$C_{29\to 31}DS_{38\to 43}C_{29\to 31}$(2)	7	–	Lam	H,M,N	168
$S_{25\to 50}C_{50\to 75}$(2)	28–30	–	–		
$I_{50\to 70}C_{30\to 50}$(3)	30–56	–	–	D,G,O	169
$S_{20\to 25}I_{50\to 70}C_{20\to 25}$(3)	56–82	–	–		
$S_{15\to 25}B_{50\to 70}C_{15\to 25}$(3)	55–81	–	–		
EOBO(2)	6–7	Gyr/Lam	Lam	A,B	170
$EO_{50}B_{50}$	11	Lam	Lam	C,M,P	171 172
$EO_{50}B_{50}/B^3$	14	Lam/Cyl/Sph	Lam(Br)/Cyl/ Sph	A,C–E	87
$E_{18\to 59}VCH_{41\to 82}$(5)	27	Sph	Sph	A,C	42
$EO_{6\to 79}I_{21\to 94}$(25)	5–84	Dis/Sph/Cyl/ HPL/Gyr/Lam	Lam	A,D G,K	173
$EO_{50}S_{50}/S^5$	23	Cyl	Cyl	A–C,J	174
$EO_{3\to 33}S_{33\to 48}I_{34\to 49}$(10)	19–30	Dis/Lam/PLS/ C-S Cyl/SPLS	–	A,C,G	120
$EO_{51}B_{49}$	11	Lam	Lam	A,C–E	175
$S_{35}B_{15}C_{50}$	150	Cyl/Lam	Cyl/Lam/Stes	C,E	125
$S_{10\to 63}EO_{4\to 37}C_{21\to 86}$(5)	24–150	–	–		
$S_{39}EO_{61}$	46	–	–		
$EO_{11}C_{78}EO_{11}$	71	–	–	B,D	30
$C_{10\to 30}EO_{4\to 37}S_{6\to 34}$	25–175	–	–		
$EO_{4\to 37}C_{10\to 30}$(4)					
$S_{14}EP_{57\to 64}E_{22\to 29}$(2)	109–119	Cyl	Cyl	A,C D,H	126
$C_{21\to 45}DTC_{55\to 73}$(6)	18–55	–	–	D,O	176
$E_{45}aP_{55}$	100	Lam	Lam	D,C	177
$EO_{39}S_{61}$	28	HPL	HPL	A–C,G	38
EOB	7–25	–	Lam/Sph	H	178
$B_{11\to 24}I_{39\to 70}EO_{19\to 42}$(4)	68–135	Cyl/Lam	Cyl/Lam		29
$E_{11\to 24}EP_{40\to 71}EO_{18\to 41}$(6)	67–138	Lam	Lam	A–D,B	100 101 119
$S_{39\to 81}EO_{19\to 61}$(2)	19–46	–	–		
$S_{27}C_{73}$	81	–	–		29
$S_{23\to 62}B_{21\to 52}C_{11\to 56}$(5)	62–110	C-S Cyl/Lam	C-S Cyl/Lam		
$S_{23\to 62}E_{21\to 52}C_{11\to 56}$(5)	62–110	Cyl/Lam	Cyl/Lam		

Table 1 (continued)

System [a]	M_n (kg mol^{-1})	Melt morphology [c]	Morphology [c]	Method [d]	Ref.
$EO_{39}S_{61}$	28	HPL	HPL	A,B	39
$EO_{55}BO_{45}/BO^{1\to2}(9)$	6–12	Lam/Cyl/ Sph	Lam/Cyl/ Sph	A,D,E	179
$EO_{46}B_{54}$	8	Lam	Lam	A,H	180
$EO_{39}{}^{4}I_{61}{}^{5}$	10	Cyl/Gyr	Lam(T)	A,D	181
$C_{3\to17}B_{83\to97}(4)$	100	Sph	Sph	A,D,B	22
$E_{14}SEB_{86}(5)$	35–74	Sph/Cyl	Dist(Br)/Sph	A–D	21 91
$EG_{34\to70}B_{30\to66}(4)$	8–16	Lam/Cyl	Cyl/Lam	A,D,E	113
$CHE_{35.5\to38.5}E_{23\to29}$ $CHE_{35.5\to38.5}(6)$	40–107	Cyl	Cyl	A,C D,H	182
$E_{26\to27}MB_{73\to74}(4)$	35[b]–8[b]	Dis/Cyl	Lam/Lam(Br)/ Cyl/Cyl(T)	A,B C,D,G	10 21
$E_{48}aP_{52}$	113[b]	Lam	Lam	A,C D,Q,R	26
$EO^{1}S^{3}$	4	Dis	Lam/ HPL	A,B,G H,R–T	183 184
$S_{29}EO_{42}S_{29}$	14	Lam	Lam/Lam(Br)	A,B,D	40
$EO_{21\to66}BO_{34\to79}(16)$	2–25	NC/Dis/ Sph/Cyl/Lam	Lam(Br)		96
$EO_{13\to28}BO_{36\to74}EO_{13\to28}(13)$	6–15	Cyl/Gyr/Lam	Lam(Br)	A,D,E	191
$BO_{12\to38}EO_{24\to66}BO_{12\to38}(15)$	5–13	Dis/Sph/Cyl/ Gyr/Lam	Lam(Br)		
$EO_{30\to43}BO_{14\to40}EO_{30\to43}(10)$	2–10	–	–	U	192
$S_{8\to33}EP_{37\to76}E_{11\to30}(3)$	115–121	–	Cyl	A,C D,H	185
$S_{54}C_{46}$	18	Lam	Lam	H,D	19
$EO_{20\to70}D_{30\to80}(2)$	18	–	–	D	109
$S_{54}C_{46}$	19	–	–	H	186
$EO_{17}B_{83}$	25	Sph	Sph	A,B,D	85
$S_{13}EP_{71}E_{16}$ $S_{76}E_{24}$	103 50	Cyl Cyl	Cyl Cyl	A,C,D	187
$D_{23\to77}C_{77\to23}(3)$	11–35	–	–	A,B D,E	107
$D_{23}C_{77}$	35	–	Lam	H	103
$EO_{17}B_{83}$	33	Sph	RL or EP	A,C	188
$EO_{21}B_{79}/B_{12\to17}{}^{1\to3}(4)$	11–26	Sph	E	A,C	189
$EO_{37\to48}S_{52\to63}(4)$	18–118	Lam	Lam, DG, C	A,B,D	193 194
$EO_{37}BO_{63}/B^{2}$ $S_{30}EO_{40}S_{30}$	7 16	Sph Lam	Sph Lam(Br)	A,D	195

Table 1 (continued)

System [a]	M_n (kg mol^{-1})	Melt morphology [c]	Morphology [c]	Method [d]	Ref.
$E_{48}EO_{52}$	2	Lam	Lam	A,B D,M	197
$C_{42\rightarrow 69}E_{31\rightarrow 58}(3)$	8–18	Lam/Cyl	Lam	A,C,D	198
$LLA_{32\rightarrow 71}EG_{68\rightarrow 29}(3)$	7–17	–	Stes	B,D, E,V	199
$S_{41}LLA_{59}$	35	Lam	Lam	A,B, C,D	196
$LLA_{6\rightarrow 37}EO_{89\rightarrow 26}LLA_{6\rightarrow 37}(4)$	40–51	Hom	Stes	A,B, D,E	200
$LLA_{44}C_{56}$	21	Hom	Stes	A,B, D,E	201
$LLA_{60}C_{40}$	25	Hom/Lam	Stes	A,B, D,E	201
$EO_{19}C_{81}$	27 [b]	–	Stes	B,D,E	202

[a] System: aP(atactic polypropylene); B(Polybutadiene); BO(Poly(butylene oxide)); C(Polycaprolactone); Ch(Cyclohexane); CHE(Poly(cyclohexyl ethylene)); D(Polydioxanone); DC(Dodecanoic acid); DS(Poly(dimethyl siloxane)); DTC(2,2-dimethyltrimethylene carbonate); E(Polyethylene); EE(Poly(ethyl ethylene)); EG(Poly(ethylene glycol)); EI_{br} (Branched Poly(ethylene imine)); EI_{cy}(Cyclic Poly(ethylene imine)); EI_{li}(Linear Poly(ethylene imine)); EO(Poly(ethylene oxide)); EP(Poly(ethylene-alt-propylene)); ET(Poly(ethylene terephthalate)); hhP(head to head polypropylene); I(Polyisoprene); LLA(Polylactide); MB(Poly(1,3 methyl-1-butene)); MMA(Poly(methyl methacrylate)); S(Polystyrene); SEB (Random Terpolymer styrene-ethylene-butene); THF(Poly(tetrahydrofuran)); VCH(Poly-(vinyl cyclohexane)); In those cases where several compositions of the same block copolymer were prepared, the composition range is given in the subscripts, and the number in brackets immediately following the block copolymer notation indicates the number of compositions. Superscripts indicate molecular weight in kg mol^{-1}.

[b] M_w

[c] The column *Morphology* gives the final morphology after crystallization. *Lam* lamellae, *Cyl* cylinders, *Sph* spheres, *Gyr* gyroid, *DG* double gyroid, *C-S* core–shell), *Co-Cont* co-continuous, *HPL* hexagonally perforated layer, *PLS* perforated lamellar structure, *SPLS* semiperforated lamellar structure, *PCcrys* perforated C-block crystal, *OLS* oriented lamellar structure, *UL* undulated lamellae, *Hom* homogeneous, *Inter* intermediate, *Dis* disordered, *Dist* distorted, *OS* ordered structure, *Ax* axialites, *Stes* spherulites, *E* ellipsoidal, *RL* rodlike, *EP* elongated prolate, *Br* breakout, *T* template, *In* inconclusive, *NC* not conclusive.

[d] *A* small-angle X-ray scattering, *B* wide-angle X-ray scattering, *C* transmission electron microscopy, *D* differential scanning calorimetry (*DSC*), *E* optical microscopy, *F* short-wave reflection method, *G* rheology, *H* atomic force microscopy, *I* shear wave reflection), *J* wide-angle.

X-ray diffraction, *K* dilatometry, *L* large-amplitude oscillating shear, *M* Fourier transformed IR spectroscopy, *N* X-ray photoelectron spectroscopy, *O* dynamic contact angle, *P* stress–strain measurements, *Q* interference optical microscopy, *R* small-angle light scattering, *S* small-angle neutron scattering, *T* dynamic light scattering, *U* static light scattering, *V* Raman spectroscopy.

We will start this review by considering the crystallization within isolated MD structures and its consequences for the nucleation phenomena. This is a subject that has been presented in a previous review [44], but where very recent works have made an important impact in its understanding, we feel that a unifying picture from a historical perspective is needed.

2 Homogeneous Nucleation and Fractionated Crystallization

Polymer nucleation can occur via spontaneous chain aggregation to form homogeneous nuclei. This homogeneous nucleation process requires the creation of new surfaces and, therefore, is energetically costly and occurs at typically large supercoolings. The nucleation on pre-existing surfaces is energetically favored and therefore most polymers in the bulk tend to nucleate on heterogeneities (catalyst debris, impurities and other types of heterogeneities). This process is known as heterogeneous nucleation. A special case to be considered later is that of self-nucleation where self-seeds or crystal fragments of the same polymer are generated to be used as ideal surfaces where the polymer can epitaxially nucleate. Since homogeneous nucleation would require the total absence of impurities and very large supercoolings, it is a rarely occurring phenomenon in bulk polymers.

The homogeneous nucleation phenomenon was first studied by droplet crystallization experiments performed on metals [57–60], alkanes [61] and polymers [62–66] when dispersed in inert low molecular weight media. The idea was that when the polymer in the bulk is subdivided into a number of droplets that is larger than the number of active heterogeneities present in the polymer, there should be a certain number of droplets without any active heterogeneity.

The preparation of immiscible polymer blends is another way to disperse a bulk polymer into fine droplets. It has been reported for several polymers that when they are dispersed in immiscible matrices into droplets with average sizes of around 1 μm, they usually exhibit multiple crystallization exotherms in a differential scanning calorimetry (DSC) cooling scan from the melt (at a specific rate, e.g., 10 °C min^{-1}). Frensch et al. [67] coined the term "fractionated crystallization" to indicate the difference exhibited by the bulk polymer, which crystallizes into a single exotherm, in comparison with one dispersed in a large number of droplets, whose crystallization is fractionated temperature-wise during cooling from the melt.

In order to illustrate the fractionated crystallization behavior we will present here previous results on immiscible atactic PS and isotactic polypropylene blends (iPP) [68]. The cooling behavior of PS, iPP and an 80/20 PS/iPP blend is presented in Fig. 1, as well as that of an "unmixed blend", labeled

80/20 UB (prepared using the same weight proportions of PS and iPP as in the corresponding melt mixed blend, but placing both polymers in a DSC pan separated by aluminum foil, so that no contact between the two polymers is made). The iPP homopolymer crystallizes with a single exotherm at 111 °C. The cooling DSC scan of PS only shows its T_g, at approximately 100 °C. A superposition of these two scans can be seen in the 80/20 PS/iPP UB. Figure 1 shows the crystallization of the 80/20 PS/iPP melt mixed blend as compared with that of the UB (80/20 UB). The iPP is dispersed in fine droplets of approximately 1 μm in diameter [68].

The melt mixed 80/20 PS/iPP blend displays a set of exotherms, where the amount of the iPP component that was heterogeneously nucleated is substantially reduced as indicated by the decrease of the crystallization enthalpy in the temperature region where the iPP crystallizes in bulk, i.e., at 109–111 °C (exotherm labeled A). This effect is due to the confinement of iPP into a large number of droplets. If the number of droplets of iPP as a dispersed phase is greater than the number of heterogeneities present in the system, fractionated crystallization occurs. The number of droplets for this composition is known (by scanning electron microscopy observations) to be of the order of 10^{11} particles cm^{-3} and polarized optical microscopy (POM) experiments have shown that this iPP contains approximately 9×10^6 heterogeneities cm^{-3}. In fact, it can be seen in Fig. 1 that the fractionated crystallization of the iPP compon-

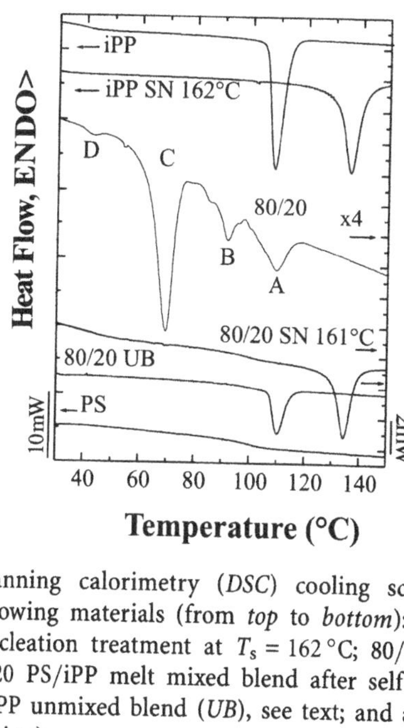

Fig. 1 Differential scanning calorimetry (*DSC*) cooling scans from the melt, at 10 °C min^{-1}, of the following materials (from *top* to *bottom*): Isotactic polypropylene (*iPP*); iPP after self-nucleation treatment at T_s = 162 °C; 80/20 polystyrene (*PS*)/iPP melt mixed blend; 80/20 PS/iPP melt mixed blend after self nucleation treatment at T_s = 161 °C; 80/20 PS/iPP unmixed blend (*UB*), see text; and atactic PS homopolymer. (From [68] with permission)

ent follows, in such a way, that the iPP crystallization occurs in at least four distinct steps, labeled A, B, C and D in decreasing temperature order [68].

The most likely explanation for the appearance of four different exotherms in the melt mixed PS/iPP 80/20 blend is the fact that when the polymer is dispersed into droplets, the content of the heterogeneities of each droplet is probably different and also the polymer may contain different types of heterogeneities which activate at different supercoolings, depending on their specific interfacial energy differences with the polymer melt. When the polymer is in the bulk, the heterogeneity with the lowest specific interfacial energy difference (e.g., heterogeneity A) will be activated at lower supercoolings and will dominate the crystallization of the polymer via secondary nucleation at the crystals created. This is the case of iPP in Fig. 1, in which there is no chance for other heterogeneities that may be activated at higher supercoolings to cause any nucleation since the polymer crystallizes at higher temperatures (it is precisely this effect that can be inhibited if the original volume of the material is divided into many small droplets). However, if the polymer also contains less active heterogeneities, which we shall term B and C, they could cause nucleation only when the polymer is dispersed into fine droplets.

For the 80/20 PS/iPP blend we could consider that only a certain droplet population contains type A heterogeneities while others may contain type B ones and so on. A number of exotherms will be generated depending on the relative supercoolings needed to activate each dominating heterogeneity within a certain droplet population. Statistically, some droplets will contain no heterogeneities at all, and in that case, homogeneous nucleation will occur. In the present case, the origin of the lowest-temperature exotherm observed (i.e., exotherm D) may be the crystallization of a small group of heterogeneity-free droplets that could have been homogeneously nucleated. Homogeneous nucleation should occur at the highest attainable supercooling for a specific volume or droplet size. For detailed studies on homogeneous nucleation and fractionated crystallization of polyolefin droplets in an immiscible matrix, the reader is referred to previous works and references therein [68–74].

The ultimate demonstration that the peculiar crystallization behavior of the 80/20 PS/iPP blend is due to the lack of highly active nuclei in every droplet is provided by self-nucleation experiments and/or by the addition of a nucleating agent [68–74]. In a self-nucleation experiment, a polymer with an initial crystalline "standard" state is heated to a given temperature, denoted the self-nucleation temperature, T_s [75]. If T_s is high enough to melt most of the polymer except for a certain number of crystal fragments, recrystallization takes place upon subsequent cooling, employing as nuclei the crystallographically "ideal" seeds which are left unmolten during heat treatment at T_s. When T_s is lower, partial melting is achieved and a large population of crystals is not melted and therefore anneals during heat treatment at T_s. Normally, three self-nucleation domains can be ascribed to crystallizable polymers depending on the self-nucleation temperature applied . In domain I or the "complete melting domain", the crystallization temperature (T_c) upon cooling from T_s remains constant and no self-nucleation can be detected. Domain II or the "self-nucleation domain" occurs when heat treatment at T_s causes a shift in crystallization temperature (during subsequent cooling from T_s) to higher temperatures with decreasing self-nucleation temperature. Finally, in domain III or the "self-nucleation and annealing domain", annealing and self-nucleation take place simultaneously.

Figure 1 shows the DSC cooling scan of iPP in the bulk after self-nucleation at a self-seeding temperature T_s of 162 °C (in domain II). The self-nucleation process provides a dramatic increase in the number of nuclei, such that bulk iPP now crystallizes at 136.2 °C after the self-nucleation process; this means with an increase of 28 °C in its peak crystallization temperature. In order to produce an equivalent self-nucleation of the iPP component in the 80/20 PS/iPP blend a T_s of 161 °C had to be employed. After the treatment at T_s, the cooling from T_s shows clearly in Fig. 1 that almost every iPP droplet can now crystallize at much higher temperatures, i.e., at 134.5 °C. Even though the fractionated crystallization has disappeared after self-nucleation, it should also be noted that the crystallization temperature in the blend case is nearly 2 °C lower than when the iPP is in the bulk; this indicates that when the polymer is in droplets the process of self-nucleation is slightly more difficult than when it is in the bulk. In the case of block copolymers when the crystallization is confined in nanoscopic spheres or cylinders it will be shown that self-nucleation is so difficult that domain II disappears.

It should be pointed out that another possible source of nucleation is the interface between the two phases under consideration. In some of the early droplet works, the authors found that the greatest supercooling needed for the crystallization of a certain droplet population was dependent on the superficial characteristics of the droplets [60, 62, 66].

Turnbull and Cech [58] analyzed the solidification of small metal droplets in sizes ranging from 10 to 300 μm and concluded that in a wide selection of metals the minimum isothermal crystallization temperature was only a function of supercooling and not of droplet size. Later, it was found that the frequency of droplet nucleation was indeed a function of not only crystallization temperature but also of droplet size, since the probability of nucleation increases with the dimension of the droplet [76]. However, for low molecular weight substances the size dependence of the homogeneous nucleation temperature is very weak [77–80].

In the case of polymer blends, the fractionated crystallization phenomenon that has been widely reported for many polymer systems can not be attributed to simple size effects. For instance, in Fig. 1, one could argue that the different exotherms originated in the crystallization of different droplet populations that have diverse average diameters. This cannot be the case, since the droplet distribution is monomodal and a smooth variation in heat evolved as the crystallization temperature is lowered would be expected for size effects only, so a single exotherm would be the result of the crystallization of such a monomodal distribution of droplets. This is indeed the case for each exotherm examined. So, within exotherm C, for example, there is a distribution of crystallization temperatures that probably reflects the droplet size distribution for that particular droplet population that contains the weakly nucleating heterogeneity C.

It has been previously shown [76] that the rate of formation of homogeneously nucleated folded chain nuclei, I, is given by

$$I = I_0 \exp\left(-\Delta F^{\pm}/kT\right) \exp\left[-\frac{32\sigma^2\sigma_e \left(T_m^0\right)^2}{kT\left(\Delta T\right)^2\left(\Delta H_v\right)^2}\right], \qquad (1)$$

where I_0 is a temperature-independent pre-exponential function that depends on molecular parameters and sometimes on crystal geometry, $\Delta F^{\pm}$ is the energy barrier for diffusion across the melt–crystal interface, k is the Boltzmann constant, T is the absolute temperature at which crystallization occurs, σ and σ_e are the lateral and fold surface free energy of the nucleus, respectively, T_m^0 is the equilibrium melting point, ΔT is the supercooling ($= T_m^0 - T$) and ΔH_v is the enthalpy of crystallization of an infinite crystal.

The freezing of droplets of a given size follows a first-order equation that can be written as [76]

$$1 - N_F/N_o = \exp\left(-k_i t\right), \qquad (2)$$

where N_F is the number of droplets frozen at time t, N_0 is the total number of droplets of size i and k_i is given by

$$k_i = v_i I, \qquad (3)$$

where v_i is the volume of each droplet of size i. A rate dependence on droplet volume is therefore expected. It has been argued [76] that since the supercooling for homogeneous nucleation is usually large, the growth rate will also be very large; therefore, once a nucleus is formed in a drop, the drop will instantaneously crystallize. Using this argument, we can equate the rate of droplet solidification to the rate of nucleus injection.

For heterogeneous nucleation the equivalent nucleation rate is given by [76]

$$I = I_0 \exp\left(-\Delta F^{\pm}/kT\right) \exp\left[-\frac{16\sigma\sigma_e\left(\Delta\sigma\right)\left(T_m^0\right)^2}{kT\left(\Delta T\right)^2\left(\Delta H_v\right)^2}\right], \qquad (4)$$

where $\Delta\sigma$ is a complex surface free-energy term that takes into account the differences in interfacial energy between the heterogeneity causing the nucleation, the crystal and the melt. Comparing Eqs. 1 and 4, we can be appreciate that as long as $\Delta\sigma < 2\sigma$ at a given temperature, the heterogeneous nucleation process will dominate and homogenous nucleation will always be absent in this case. The temperature dependence of the nucleation rate in the homogeneous case should therefore be larger than in the heterogeneous case. Apart from this fact, in both cases a plot of I versus $1/[T(\Delta T)^2]$ should yield a straight line. Therefore, it is not an easy task to demonstrate that the process is homogeneous with absolute certainty from nucleation rate measurements as a function of temperature. The values of the product of superficial energies can be calculated but they are difficult to compare with theoretical estimates or with values obtained by other techniques since the reported range can be quite large for a given polymer.

Turnbull [60] was able to show in the case of mercury droplets (2–8 μm) that when they are coated with mercury laureate, they solidify isothermally at rates that are proportional to the droplet volume at maximum supercooling (isothermal crystallization temperatures from – 117 to – 119 °C). The isothermal freezing kinetics was satisfactorily described by the homogeneous nucleation theory for condensed systems [81]. When the droplets were coated with mercury acetate, they crystallized isothermally at higher temperatures; the kinetics of solidification was determined in a temperature range from – 85.01 to – 86.08 °C. In this case, the frequency of droplet freezing was proportional to droplet area and was in good agreement with a heterogeneous nucleation theory that describes the formation of crystal embryos on the surface of a nucleation catalyst [82]. In this case, it was clearly demonstrated that the surface of the droplet was causing the nucleation effect. It is remarkable that in both cases, the volume homogeneous nucleation and the surface heterogeneous nucleation, the rate of crystal nucleation followed first-order reaction kinetics described at constant temperature for a given dispersion by a single value of nucleation frequency/volume or area. Therefore, only finding a first-order nucleation rate is not a demonstration that volume homogeneous nucleation has been obtained since a surface nucleation effects may also yield similar kinetics.

An ingenious new method of studying the crystallization of finely dispersed poly(ethylene oxide) (PEO) droplets was recently reported by Massa et al. [83]. They produced an ensemble of impurity-free PEO droplets by dewetting a spin-coated thin film on an unfavorable PS substrate. The dewetting process is capable of cleaning the material, leaving on the PS substrate the contaminants that can act as heterogeneities. The size of the PEO droplets produced by this technique can be controlled by varying the thickness of the deposited film before dewetting. In the example presented in their paper, the authors produced a PEO ($M_w = 27\,000$ and $M_w/M_n = 1.09$) droplet distribution with diameters ranging from 8 to 16 μm. They determined by POM the fraction of droplets that were able to crystallize as a function of temperature during controlled cooling (0.4 or 0.1 °C min^{-1}) or as a function of time during isothermal runs. In the case of controlled cooling from the melt, they found that the droplets crystallized from – 2 until – 7 °C. For PEO in the bulk or in the case of droplets containing heterogeneities, the crystallization range employing identical conditions was 56–46 °C. In order to demonstrate that the clean droplet dispersion was nucleating homogeneously, they employed correlation plots for two consecutive runs where they followed the solidification of the same group of droplets. In the case of heterogeneous nucleation, a clear correlation was found between the crystallization temperatures of the first and the second run, indicating that the defect that induces nucleation is determining the activation energy. In contrast, for homogeneous nucleation, the data were clustered around the temperature at which the nucleation rate was maximum, – 5 °C, since the activation energy is a function of PEO itself and is therefore identical within each droplet provided the size is the same.

Figure 2 presents results from Massa and Dalnoki-Veress [84] where the fraction of solidified droplets as a function of temperature is represented for an ensemble of clean PEO droplets obtained by dewetting. An optical micrograph taken at a specific temperature (T_c = – 2.6 °C) indicates that the crystallized droplets are easily recognizable by their birefringence as compared with dark amorphous droplets. Performing isothermal experiments, they monitored the crystallization of droplets of several sizes. Then, they fitted an equation similar to Eq. 2 to their experimental data for different droplet diameters. They found that, as expected, the nucleation rate was faster for larger droplets and by plotting the time constant associated with the nucleation event (i.e., $\tau = 1/k_i$) as a function of droplet radius in a semilog plot, a scaling of $\tau \propto R^{-3}$ was obtained. This result indicates that homogeneous bulk nucleation is present in the PEO droplets and that the surface is not responsible for the nucleation observed.

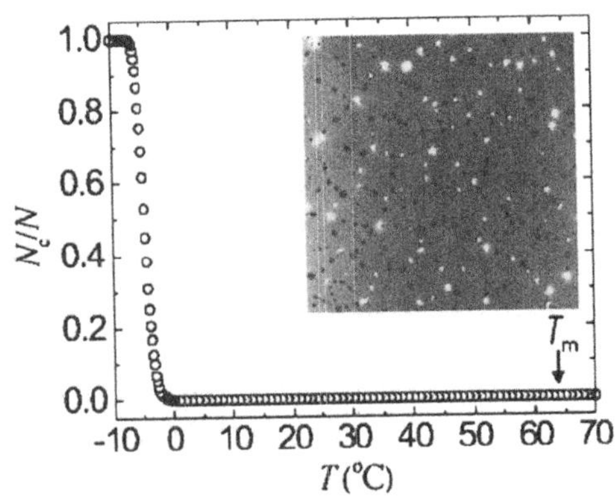

Fig. 2 Optical microscopy image of a small section of a poly(ethylene oxide) (*PEO*) droplet dispersion sample, see text (1000-mm wide) obtained at T_c = – 2.6 °C. Amorphous droplets appear dark and semicrystalline droplets appear white under nearly crossed polarizers. The plot shows the fraction of crystallized droplets as a function of temperature upon cooling (0.4 °C min^{-1}) for homogeneous nucleation. (Reprinted with permission from [84]. Copyright 2004 by the American Physical Society)

The effect of particle size on the homogeneous nucleation temperature of droplets has been examined for a number of low molecular weight materials and although a clear trend of decreasing crystallization temperature of the droplets as the droplet diameter decreases has been found, the crystallization temperature only varies within 1 or 2 °C in the materials and droplet size range examined so far. For example, water droplets trapped within liposomes freeze at decreasing temperatures [77]; when the particle size decreases from 0.49 to 0.14 μm the crystallization temperature decreases from – 40.6 to – 41.7 °C (these temperatures were determined by DSC at a cooling rate of 10 °C min^{-1}). In the case of narrowly dispersed microdroplets produced by a miniemulsion, droplets of hexadecane ranging in diameter from 410 to 136 nm have been produced [79]. The droplets can homogeneously nucleate and their crystallization temperatures decreased monotonically from – 3.2 to – 5.0 °C as the droplet size decreased (these temperatures were determined by DSC at a cooling rate of 5 °C min^{-1}). A similar result was found for NaCl solution droplets ranging from 330–109 nm, whose crystallization temperatures decreased from – 44.6 to – 46.6 °C. A series of alkane droplets prepared by mini-emulsion have yielded very similar results [79].

Figure 3 shows a plot of the volume normalized nucleation time constant as a function of isothermal crystallization temperature for PEO droplets, taken from the work of Massa and Kalnoki-Veress [84]. As expected, droplets of different volumes have the same value of τV. The inset in Fig. 3 is a plot consistent with classical nucleation theory (see Eqs. 1, 4); only the last four data points correspond to the work of Massa and Kalnoki-Veress. The first

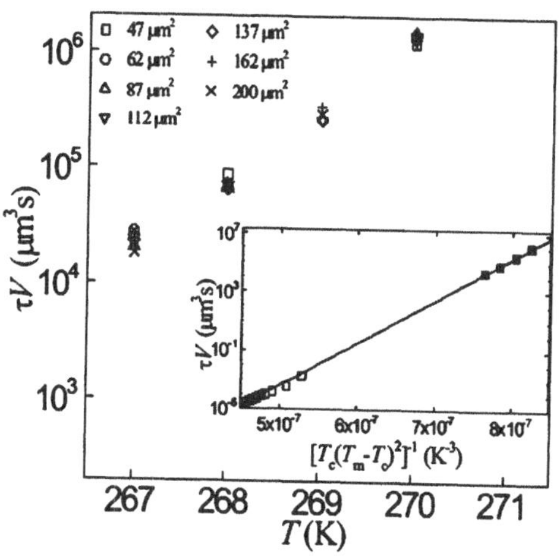

Fig. 3 Semi-logarithmic plot of the volume-normalised time constant, τV, as a function of temperature. The data shows a linear dependence when τV is plotted as a function of $1/[T_c(T_m - T_c)^2]$ as is expected from classical nucleation theory (see inset). (Reprinted with permission from [84]. Copyright 2004 by the American Physical Society)

group of data points correspond to experiments of Röttele et al. [85], who studied by atomic force microscopy the crystallization of PEO within a PB-*b*-PEO block copolymer (with 17% PEO and 4300 g mol^{-1} for the PEO block). The PEO was confined into 12-nm-diameter spherical MDs in a PB matrix. The isothermal crystallization temperature range for these much smaller PEO spheres encompasses – 31 to – 24 °C, while the data of Massa and Kalnoki-Veress are for much larger PEO droplets of approximately 8–16 μm in diameter that were isothermally crystallized in the range – 6 to – 3 °C. In both cases the authors demonstrated that the nucleation was homogeneous. This is probably the largest homogeneous nucleation temperature range ever reported for any material crystallized into droplets. The reason for such behavior is probably the large differences in droplet volumes, which span a range from 905 nm^3 for the 12-nm-diameter spheres to 268–536 × 10^9 nm^3 in the case of the larger droplets (diameter range 8–16 × 10^3 nm), i.e., a factor of approximately 10^9 difference in volume or 10^3 in diameter. Such a large difference in droplet size produced a maximum difference in the homogeneous crystallization temperature of 28 °C, an unusually large variation.

As already discussed, homogeneous nucleation is a process that may be encountered when the material is finely dispersed into droplets or isolated MDs, as long as the number of MDs outnumbers all active heterogeneities present in the polymer by a large factor. The difference with heterogeneous nucleation may not be easily established by the temperature dependence of the nucleation rate unless firm knowledge of superficial energies is independently gained. Homogeneous nucleation is a process that should occur at the largest possible supercooling, although the temperature range has been found to vary depending on the specific material employed (Sect. 3). It is also noteworthy that homogeneous nucleation can be described by first-order kinetics. However, possible surface nucleation effects that could also be described by first-order kinetics must be taken into consideration. One way to distinguish between surface and bulk homogeneous nucleation is to determine the scaling of the time constant, associated with the nucleation event, with the particle diameter in the case of droplets. Another important factor is that since the nucleation is time-dependent, care should be taken upon comparing dynamic and isothermal crystallization temperatures before attempting to interpret them as arising from homogeneously nucleated crystals. Finally the volume of the droplets or the MD should also be considered, since it can be very important for specific materials.

3 Homogeneous Nucleation and Fractionated Crystallization in Block Copolymer Microdomains

Several factors contribute to make block copolymers with crystallizable blocks an ideal system to study homogeneous nucleation: the purity involved in the synthesis, the size of the MD which can easily reach the nanoscopic scale and the nature of the interphases that are present in strongly segregated systems, which could lead to less interference from interphase nucleation than the interface between dispersed droplets and the matrix in immiscible blends. Nevertheless, this last possibility (i.e., surface nucleation) may also occur in block copolymers and needs considering. On the other hand, the connectivity of the chains through covalent bonds and the variable segregation strength (which depends on χN) means that if the MDs are not well defined, the possibility of percolation and therefore of the spread of heterogeneous nucleation by secondary nucleation can complicate the behavior.

Since PEO is a polymer that has been widely used as a crystallizable block copolymer component, and there is also a wide range of data on PEO droplets produced in a wide range of sizes, we would like to use PEO as an example to illustrate the great diversity of literature reports on its crystallization temperature and kinetics as a function of morphology. One interesting aspect about PEO is that its glass-transition temperature (T_g) is located, depending on the sample molecular weight, in the range – 60 to – 50 °C (as determined by DSC with a heating rate of 10 °C min^{-1}), and the homogeneous nucleation temperature in the case of nanoscopic spheres has been reported to be very close to T_g [29].

Figure 4 compiles dimension and crystallization temperature data that have been reported for PEO spheres, in a very wide range of sizes. Only

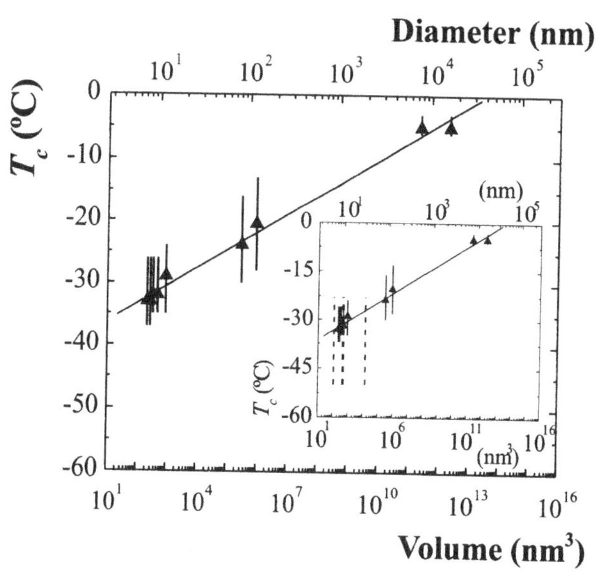

Fig. 4 Reported dynamic crystallization temperatures for PEO isolated spheres as a function of their volume (*bottom x-axis*) and diameter (*top x-axis*). The *inset* shows dynamic (*symbols with vertical bars*) and isothermal crystallization (*vertical dashed bars*) temperature ranges for PEO spheres, see text

references where the PEO spheres have most probably been homogeneously nucleated were chosen to prepare this figure. The biggest spheres correspond to those prepared by dewetting by Massa et al. [83], the intermediate-size PEO spheres (diameters of approximately 100 nm) were prepared by miniemulsion polymerization [86]. The smallest spheres correspond to nanodroplets obtained within block copolymers or in blends of block copolymers with a homopolymer where spherical morphologies have been obtained (Table 2).

The triangles in Fig. 4 correspond to peak crystallization temperatures (or the temperature at which the crystallization rate was a maximum) obtained during cooling PEO from the melt at rates ranging from 5 to 0.4 °C min^{-1}; these will be referred to as dynamic crystallization temperatures. In references where DSC cooling scans are reported, we have measured the crystallization onset and final temperatures and plotted them with a vertical line as a temperature range around the triangles in Fig. 4. On the other hand, there have been several reports of isothermal crystallization of PEO spheres and the experiments reported therein were usually performed for a wide range of crystallization temperatures; all of these temperatures are reported in the inset of Fig. 4 using vertical dashed lines to encompass the reported isothermal temperature range (both isothermal and dynamic crystallization temperatures are reported in the inset, only the data reported with triangles

Table 2 Data corresponding to poly(ethylene oxide) (*PEO*) spheres prepared by different methods (see Fig. 4)

Diameter (nm)	System	Reference	T_c range
6.0	$EO_{17}B_{20}/PB_{63}$	188	- 50 → - 23 i
7.5	$EO_{17}B_{20}PB_{63}$	87, 92, 189	- 37 → - 26 d
8.1	$EO_{17}B_{20}/PB_{63}$	87, 92, 189	- 37 → - 26 d
8.7	$EO_{17}B_{71}/PB_{12}$	87, 92, 189	- 35 → - 26 d
9.4	$EO_{17}B_{71}/PB_{12}$	87, 92, 189	- 50 → - 23 i
9.7	$EO_{17}B_{20}/PB_{63}$	87, 92, 189	- 35 → - 26 d
10	$EO_{17}B_{83}$	188	- 50 → - 23 i
10	$EO_{17}B_{20}/PB_{63}$	87, 92, 189	- 50 → - 23 i
12	$B_{83}EO_{17}$	23, 85	- 35 → - 24 d
30	EO_wB_y	188	- 50 → - 23 i
83	PEO nanodroplets	86	- 30 → - 15 d
120	PEO nanodroplets	86	- 28 → - 13 d
7600	PEO droplets	83, 84	- 6 → - 3 d
16000	PEO droplets	83, 84	- 6 → - 3 d

i: isothermal crystallization data, d: dynamic crystallization data correspond to dynamic crystallization conditions). For the purpose of finding a correlation between crystallization temperature and sphere volume or diameter, only the values measured under dynamic conditions (peak crystallization temperatures or the temperature at which the crystallization rate was a maximum during cooling from the melt) were taken into account. The fittings to these dynamic crystallization temperatures are very good:

$$T_c = \left[-41.8 + 2.89 \log\left(v_d\right)\right] , \tag{5}$$

$$T_c = \left[-39.9 + 8.9 \log\left(d\right)\right] , \tag{6}$$

where v_d is the droplet volume and d the droplet diameter. The correlation coefficient for these fittings was 0.995 in both cases, in spite of the variation in scanning rates employed by different authors. The data employed to construct Fig. 4 can be found along with the corresponding references in Table 2. It is clear that in the case of PEO a correlation exists between droplet size and the homogeneous nucleation temperature, as was previously pointed out by Massa and Kalnoki-Veress [84] and by Chen et al. [87] since the probability of nucleation and the nucleation rate (Eq. 3) depend on sample volume. The variations of T_c are in this case quite large compared with those reported for low molecular weight compounds for which droplet size was varied.

Another interesting implication of the data compiled in Fig. 4 is that when T_c is plotted as a function of the inverse of the sphere diameter, the limit where the sphere diameter tends to infinity should correspond to the crystallization temperature of homogeneously nucleated PEO in the bulk. In other words, this would be the maximum temperature at which homogeneous nucleation for PEO could ever be observed, for very large heterogeneity-free PEO phases or even bulk PEO. Such a temperature depends on the fitting expression employed; however, it should correspond to a crystallization temperature close to – 5 °C or between – 10 and 0 °C.

The overall isothermal crystallization kinetics of polymers can be described by the Avrami equation [88–90]:

$$1 - v_c = \exp(-Kt^n) , \tag{7}$$

where n is the Avrami index, a parameter that indicates the dimensionality of growth together with nucleation type, K is the overall relative transformation rate constant (its value includes contributions from both nucleation and growth) and v_c is the relative volume crystalline fraction (its value ranging from 0 to 1; therefore, it is a normalized value with respect to the maximum crystallinity achieved during the crystallization time employed). When $n = 1$, the Avrami equation reduces to a first-order equation of the type presented before to rationalize the freezing of polymer droplets (Eq. 2).

Loo et al. [21, 42, 44, 91] reported a simple exponential decay ($n = 1$) for polyethylene-*b*-poly(styrene-*ran*-ethylene-*ran*-butene) [PE-*b*-poly(SEB)] diblock copolymers in which the PE blocks form isolated spheres or cylinders. They have also explored the influence of the confinement level on the crystallization kinetics of PE-*b*-poly(vinylcyclohexane) (PE-*b*-PVCH) diblock copolymers [42]. They concluded that the crystallization kinetics strongly reflects the connectivity of the PE MDs in such way that homogeneous nucleation and first-order kinetics were obtained for spheres or cylinders of PE, whereas conventional sigmoidal kinetics (i.e., $n > 1$) is obtained for the highly interconnected gyroid structure and in some cases for lamellar diblocks which have defects that cause certain interconnectivity. They claim that the first-order kinetics implies that crystallization is isolated within individual MDs, resulting in a crystallization rate that is proportional to the fraction of material not yet crystallized, i.e., that the crystal growth must be essentially instantaneous within spherical or cylindrical domains and homogeneous nucleation is the rate-determining step in the crystallization. Such arguments had been employed previously when analyzing the freezing of polymer droplets [76]. This assumption is based on the fact that the free-energy barrier for homogeneous nucleation is much higher that that associated with crystal growth. Therefore, the n exponent depends only on the volume of the crystallizable MD and not on the MD dimensionality or shape. Such dependence was initially proposed by Turnbull et al. [60–63] when they studied the crystallization kinetics of droplets. Similar first-order kinetics for PEO spheres was obtained by Chen et al. [92], Massa and Kalnoki-Veress [84] and Reiter et al. [23]. Zhu et al. [37] also found values of n of around 1 for PEO cylinders, but they gave a different explanation for their results. They attributed the low n values to a hard confinement that restricts the crystal growth of PEO-*b*-PS diblocks to one dimension assuming that the nucleation is athermal. Their crystallization temperatures in the cases where n was 1 or lower were also too high (i.e., 22–30 °C) as to be associated with a previous homogeneous nucleation process.

Lotz and Kovacs [93] reported $n = 0.5$ many years ago when their PS-*b*-PEO diblock copolymers crystallized at very large supercoolings (i.e., homogeneous nucleation was assumed to take place). However, the origin of this result was not explained.

Balsamo et al. [94] in a recent contribution considered that the Avrami exponent is given by the addition of two terms:

$$n = n_{\mathrm{n}} + n_{\mathrm{gd}} , \tag{8}$$

where n_{n} is the fraction of the exponent related to nucleation and n_{gd} that related to growth dimensionality. Using the conventional Avrami analysis that takes into account whether the nucleation is instantaneous or sporadic (n_{n} would be 0 or 1, respectively) and the dimensionality of the growth (n_{gd} can have values from 1 to 3 for one-dimensional to three-dimensional growth), values less than 1 cannot normally be explained, because 1 would be obtained when the nucleation is athermal (or instantaneous) and the growth is one-dimensional. However, if we follow the argument that growth is so fast that the crystallization kinetics is completely dominated by the nucleation process (a consideration that would be equivalent to considering $n_{gd} = 0$ in Eq. 8), then a value of $n < 1$ can only imply that nucleation is not completely sporadic, but that it is somewhere in between sporadic and instantaneous. In fact, in most cases polymer nucleation is not purely instantaneous or purely sporadic but ranges between the two extremes, and this is considered to be one of the reasons why integer Avrami indexes are rarely encountered [95].

Figure 5 and Table 3 present a compilation of literature data corresponding to the crystallization of PEO for block copolymers with a PEO block, or blends of such block copolymers and a homopolymer. The data comprise both dynamic and isothermal crystallization temperatures for all types of morphology reported, i.e., spheres, cylinders and lamellae, regardless of whether the crystallization proceeds via homogeneous or heterogeneous nucleation. Bulk PEO is also included. In some cases, the authors performed crystallization kinetics and obtained values of the Avrami index of 1 or lower and this is also represented in Fig. 5 (with filled bars), since this could indicate that homogeneous nucleation was encountered. The horizontal lines shown in Fig. 5 encompass a temperature range (between − 10 and 0 °C, as extrapolated previously) that represents the approximately maximum temperature that could be associated with homogeneous nucleation according to the extrapolation procedure performed with the data of Fig. 4 and explained earlier. A general rough trend of increasing crystallization temperature in the order

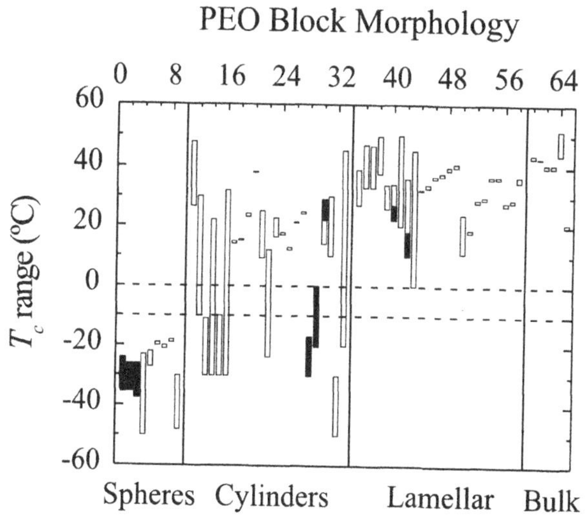

Fig. 5 Schematic plot showing reported crystallization temperatures for PEO in the bulk and as a component of block copolymers of varying compositions. The morphology of the PEO block is indicated on the *x*-axis. The *filled bars* are for data where isothermal crystallization measurements were performed and Avrami indexes of 1 or less were reported. The *horizontal lines* indicate the maximum temperature range that can be associated with PEO homogeneous nucleation, see text

Table 3 Block copolymer systems that incorporate PEO as a component (see Fig. 5 and text)

Bar	Reference	System	Bar	Reference	System
Spheres			Lamellae		
1	23, 85	$B_{83}EO_{17}$	35	36	$EO_{49}S_{51}$
2	87, 92, 189	$EO_{17}B_{20}/PB_{63}$	36	113	$EO_{57}B_{43}$
3	87, 92, 189	$EO_{17}B_{20}/PB_{63}$	37	113	$EO_{51}B_{49}$
4	188	$EO_{17}B_{83}$	38	113	$EO_{33.5}B_{33}EO_{33.5}$
5	87, 92, 189	$EO_{17}B_{20}/PB_{63}$	39	113	$EO_{21}B_{58}EO_{21}$
6	96	$EO_{21}BO_{79}$	40	37	$EO_{47}S_{53}$
7	96	$BO_{38}EO_{24}BO_{38}$	41	175, 172	$EO_{53}B_{47}$
8	96	$BO_{37.5}EO_{25}BO_{37.5}$	42	40	$S_{18.5}EO_{63}S_{18.5}$
9	30, 99, 124	$S_{81}EO_{19}$	43	29, 100, 101, 119	$B_{19}I_{39}EO_{42}$
Cylinders			44	96	$EO_{50}BO_{50}$
11	113	$EO_{34}B_{66}$	45	96	$EO_{51}BO_{49}$
12	38	$EO_{36}S_{61}$	46	96	$EO_{66}BO_{34}$
13	29, 100	$B_{24}I_{56}EO_{20}$	47	96	$EO_{52}BO_{48}$
14	29, 100	$B_{11}I_{70}EO_{19}$	48	96	$EO_{66}BO_{34}$
15	29, 100	$E_{24}EP_{57}EO_{19}$	49	96	$EO_{52}BO_{48}$
16	29, 100	$E_{11}EP_{71}EO_{18}$	50	96	$EO_{34}BO_{32}EO_{34}$
17	96	$EO_{29}BO_{71}$	51	96	$BO_{33}EO_{34}BO_{33}$
18	96	$EO_{27}BO_{73}$	52	96	$EO_{34}BO_{32}EO_{34}$
19	96	$EO_{36}BO_{64}$	53	96	$EO_{26}BO_{48}EO_{26}$
20	96	$EO_{37}BO_{63}$	54	96	$EO_{35}BO_{30}EO_{35}$
21	96	$EO_{19}BO_{62}EO_{19}$	55	96	$EO_{38}BO_{24}EO_{38}$
22	96	$EO_{18}BO_{39}EO_{43}$	56	96	$EO_{32}BO_{36}EO_{32}$
23	96	$EO_{27}BO_{46}EO_{27}$	57	96	$BO_{17}EO_{66}BO_{17}$
24	96	$EO_{25}BO_{50}EO_{25}$	58	96	$BO_{18}EO_{64}BO_{18}$
25	96	$EO_{23}BO_{54}EO_{23}$			
26	96	$BO_{32}EO_{36}BO_{32}$	Bulk		
27	96	$BO_{32}EO_{36}BO_{32}$	60	190	–
28	87, 92, 189	$EO_{31}B_{6}/PB_{63}$	61	86	–
29	87, 92, 189	$EO_{22}B_{15}/PB_{63}$	62	87, 92, 189	–
30	37	$EO_{32}S_{53}/PS_{15}$	63	87	–
31	37	$EO_{32}S_{53}/PS_{15}$	64	113	–
32	30, 99, 124	$S_{63}EO_{16}CL_{21}$	65	99	–
33	29, 100, 101, 119	$E_{11\rightarrow 24}EP_{40\rightarrow 71}EO_{18\rightarrow 41}$			

The bar number refers to the bar in Fig. 5

spheres, cylinders, lamellae, bulk PEO can be observed that is consistent with the increase in PEO volume.

There are many peculiarities associated with the data reported in Fig. 5, in the sense that some values do not fit the general trend or expectation based on the concepts outlined in this section. Some of these will be described briefly in what follows.

We have represented schematically in Fig. 5, the maximum temperature range that can be associated with homogeneous nucleation temperatures for PEO. Some data, where Avrami indexes of 1 or lower have been reported, have crystallization temperatures that fall above this range, so they should not be associated with homogeneous nucleation. Another origin for the low Avrami indexes may be involved in these cases.

It is clear from the previous discussion that a proportionality between crystallization temperature and the volume of the crystallizing phase has been found for PEO in many cases (Fig. 4), including AB and ABA diblock copolymers; however, there have been reports of exceptions to this trend and they have also been included in the data compilation of Fig. 5, in particular the extensive data reported by Xu et al. [96]. In the case of ABC triblock copolymers a different behavior has also been reported and will be analyzed in detail in Sect. 5.

Xu et al. prepared a series of PEO-*b*-poly(oxybutylene), PEO-*b*-PBO, diblock copolymers as well as PEO-*b*-PBO-*b*-PEO and PBO-*b*-PEO-*b*-PBO triblock copolymers in a very wide composition range. In these block copolymers, only the PEO block crystallizes. They reported that for PEO-*b*-PBO diblock copolymers with "long" PEO blocks (i.e., with a molecular mass of 4840 g mol^{-1} or higher), the crystallization temperature during cooling at 10 °C min^{-1} is independent of both the volume fraction and the morphology and is always quite high (i.e., 37–40 °C, in the same order of magnitude as the crystallization temperature of bulk PEO). This is surprising since in most cases reported in Tables 2 and 3 for PEO spheres within diblock copolymers for instance, the PEO has a molecular mass equal to or higher than 4840 g mol^{-1} and the composition effect on T_c seems to be quite clear. On the other hand, the copolymers reported in Fig. 4 are based on PEO-*b*-PS and PEO-*b*-PB or blends of PEO-*b*-PB/PB, where the segregation strength is higher than in the PEO-*b*-PBO system. Furthermore, Xu et al. [96], also reported that for "short" PEO blocks (i.e., with a molecular mass lower than 4840 g mol^{-1}) within PEO-*b*-PBO, PEO-*b*-PBO and PBO-*b*-PEO-*b*-PBO the crystallization temperature increases with both the volume fraction and the length of the PEO block. The authors claimed that the PEO block in some samples was homogeneously nucleated since its crystallization temperature "falls into the temperature range normally associated with homogeneous nucleation." This criterion, however, is not conclusive since the temperatures reported (the dynamic crystallization temperatures for the PEO block within these samples were – 19.3, – 28.2, – 21 and – 18.8 °C) are not very low for PEO small MDs (e.g., they are not that close to PEO T_g values) and therefore no direct proof for the attainment of homogeneous nucleation is provided. In fact, they have demonstrated employing small-angle X-ray scattering (SAXS) that in all these samples isothermal crystallization leads to breakout and Avrami indexes between 2 and 3. The work of Xu et al. concludes that low-temperature crystallization does not always imply that crystallization within confined MDs has occurred, as they have demonstrated in their case by SAXS. This is a valid point that must be taken into consideration if only nonisothermal DSC data are available for a particular system; however, the relatively low temperature crystallization that the authors observed upon cooling from the melt does not imply homogeneous nucleation, and may have its origin in the presence of less active nucleating heterogeneities or superficial nucleation. In Sect. 5.1 results for ABC triblocks containing PEO will be presented. In some cases similar crystallization temperatures have been reported and have been ascribed to superficial nucleation effects (see later).

When nonisothermal DSC experiments are performed at a specific cooling rate from the melt, like 10 °C min^{-1}, the observation of a crystallization exotherm for a crystallizable block copolymer component at a higher supercooling than that encountered when crystallizing the same polymer in the bulk cannot be employed as the only criterion to claim that homogeneous nucleation is encountered, unless that crystallization exotherm occurs at the maximum possible supercooling, i.e., very close to the T_g of the polymer. In the case of PEO, its T_g is located around – 50 to – 55 °C. For PEO, the dynamic crystallization of homogeneously nucleated nanodroplets has been reported in the range from – 30 to – 45 °C (peak crystallization temperatures upon cooling from the melt), i.e., quite close to T_g for small PEO MDs (Table 2). The observation of such dynamic crystallization at temperatures higher than these values may be associated with superficial nucleation phenomena or the presence of less active heterogeneities, if small-volume PEO MDs are taken into consideration.

The technique of self-nucleation can be very useful to study the nucleation and crystallization of block copolymers that are able to crystallize [29, 97–103]. Previous works have shown that domain II or the exclusive self-nucleation domain disappears for systems where the crystallizable block [PE, PEO or poly(ε-caprolactone), PCL] was strongly confined into small isolated MDs [29, 97–101]. The need for a very large number of nuclei in order to nucleate crystals in every confined MD (e.g., of the order of 10^{16} nuclei cm^{-3} in the case of confined spheres) implies that the amount of material that needs to be left unmolten is so large that domain II disappears and annealing will always occur to a fraction of the polymer when self-nucleation is finally attained at lower T_s. This is a direct result of the extremely high number density of MDs that need to be self-nucleated when the crystallizable block is confined within small isolated MDs. Although this effect has been mainly studied in ABC triblock copolymers and will be discussed in Sect. 6.3, it has also been reported in PS-*b*-PEO diblock copolymers [29, 99].

The case of PEO as a crystallizable component within block copolymers, as well as isolated droplets, is atypical in the sense that it has been studied in a wide variety of block copolymers and in the widest possible volume range if micron-size droplets are also considered (Figs. 4, 5, Tables 2, 3). Nevertheless, the observed change in crystallization temperature is particularly large, not only in comparison with low molecular weight materials, as previously mentioned, but also with other polymers.

Let us briefly examined the case of PE. A polymer like high-density PE (HDPE) has around 10^9 highly active heterogeneous nuclei per cubic centimeter in the bulk (this number does not include other type of less active heterogeneities that can possibly cause nucleation at larger supercoolings), so it may be difficult to get a large enough population of dispersed droplets that are heterogeneity free by blending with an immiscible component. Arnal and Müller [72] found in the case of PS/linear low-density PE (LLDPE) blends that fractionated crystallization exotherms in the temperature range between 67 and 70 °C were encountered for droplet sizes in the range 0.5–2 μm. Recently, Loo et al. [91] studied the crystallization of the PE block within a PE/SEB diblock copolymer constituted by hydrogenated PB (14.3 wt % PE) and an amorphous random terpolymer of SEB. The PE block crystallized within 25-nm spherical mesophases, and most of the spheres were supposed to have been homogeneously nucleated because the number of MDs was 2×10^{16} spheres cm^{-3}. The authors also found crystallization temperatures between 66 and 69 °C for the PE block, which could be somewhat similar to a LLDPE in view of its branch content (incidentally the branch content in LLDPE can complicate the analysis since T_c is a function of this parameter). This comparison shows that in the case of LLDPE (assuming similar branch contents), the temperatures that have been associated in the literature with apparent homogeneous nucleation do not seem to be as highly dependent on the volume of the crystallizing phase as in the case of PEO.

In view of the previous discussion about the homogeneous nucleation temperature in PEO, an important question that arises is why, in the case of PE whose T_g values are below - 30 °C, are the dynamic crystallization temperatures available in the literature as possibly associated with homogeneous nucleation so high, in the range of 60–70 °C. A similar situation arises with iPP, whose T_g is approximately 0 °C and the homogeneous nucleation temperature range is claimed to be around 40–50 °C. However, polymers like PEO and PCL, when they are the minor components in AB or ABC block copolymers, can exhibit crystallization temperatures upon cooling from the melt that are very close to their respective vitrification temperatures. One possible speculation could be that the true homogeneous nucleation temperature of PE and PP may not have been found yet, since surface nucleation phenomena were not entirely ruled out. Surface nucleation also occurs at higher supercoolings than heterogeneous nucleation and the temperature dependence of the nucleation rate is higher than in the heterogeneous nucleation case [60].

Furthermore, an Avrami exponent of 1 could also be obtained from surface nucleation. So far, no measurements are available in the literature of the nucleation kinetics as a function of droplet or MD size, and therefore the scaling between the relevant time constant associated with the nucleation event and the MD dimensions is unknown.

From this section we can summarize the general behavior of confined crystallizable MDs. These generalizations apply to block copolymers that are in the strong segregation regime and that can crystallize within their specific MD without breakout. When a block copolymer component crystallizes within isolated MD structures like spheres, cylinders or lamellae it may nucleate homogeneously. For homogeneous nucleation to take place, several requirements should be met:

1. The number of isolated MDs should be much larger than the usual number of active heterogeneities present in an equivalent bulk sample of the crystallizable polymer.
2. The MD structures should not contain any other type of less efficient heterogeneities present that could be activated in the absence of the heterogeneities that cause nucleation at the usual supercooling range in the bulk polymer.
3. The interphases between MDs should not be able to cause nucleation of the semicrystalline phase. If the interphase displays nucleation activity, first-order crystallization kinetics may be obtained; however, the supercooling at which dynamic crystallization from the melt occurs will not be the maximum that could be possibly attained if homogeneous nucleation were in place.
4. The MD structures must be isolated and no percolation paths should exist between them that can allow the spread of secondary nucleation throughout the crystallizable component. This condition is easier to establish in sphere-forming block copolymers. When a block copolymer forms cylinders, it is more difficult to obtain samples without a certain percentage of MD percolation and in the case of lamellae, many MDs usually percolate through grain borders. As a consequence, a mixture of heterogeneous and homogeneous nucleation can be present in the same sample and a concomitant fractionated crystallization may be seen when the polymer is cooled from the melt.

If the four conditions mentioned are satisfied, homogeneous nucleation will take place typically at very large supercoolings. The crystallization temperature during cooling from the melt will probably be a function of the MD volume, although the exact behavior may vary depending on the type of material being studied. The crystallization kinetics should follow, in principle, a first-order rate equation, such as Eq. 2, or the equivalent of an Avrami equation with Avrami indexes lower or equal to 1 (Eq. 7). Additionally, if the confined crystallizable component is subjected to a self-nucleation study, domain II will be not found (Sect. 6.2). In the case of PEO or PCL, they will exhibit during dynamic cooling from the melt a low temperature crystallization exotherm that for isolated nanodroplets or very small MDs will be located very close to their respective T_g values. More specific examples of fractionated crystallization and homogeneous crystallization will be given in the sections corresponding to ABC triblock copolymers with one or two crystallizable blocks.

4
Double-Crystalline Diblock Copolymers

There have been relatively few reports dealing with double-crystalline diblock copolymers [102–110, 197–200]. The particular case of ABC triblock copolymers with two semicrystalline blocks will be presented in a separate section. Works pertaining to one of the most studied systems PCL-*b*-PEO have already been previously reviewed [43]. Recently, probably the most comprehensive studies on double-crystalline diblock copolymer systems were performed on poly(*p*-dioxanone)-*b*-PCL diblock copolymers, PPDX-*b*-PCL, and therefore several important aspects of these works [102, 103, 107] will be summarized in this section.

The PPDX-*b*-PCL diblock copolymers were recently synthesized [111] and apart from the references already mentioned, only the contribution of Lendlein and Langer [112] deals with chemically similar materials, although structurally quite different since they employed multiblock copolymers of PPDX and PCL with very low molecular weights to prepare shape memory polymers for biomedical applications.

Figure 6 presents DSC cooling and subsequent heating scans for PPDX-*b*-PCL diblock copolymers in a wide composition range, for comparable molecular weight homopolymers and for a 50/50 PPDX/PCL blend. The blend illustrates the typical immiscible nature of the two polymers and their separate crystallization and melting. In the case of the block copolymers, previous works have suggested that they are in the weakly segregated regime since crystallization drives structure formation, such that both blocks form lamellar structures [103]. The cooling DSC scans of Fig. 6 indicate that regardless of the composition in the diblock copolymers only one crystallization exotherm is observed. Wide-angle X-ray scattering (WAXS) studies performed at room temperature after cooling from the melt indicated that both blocks had crystallized during cooling [107].

Figure 7 shows a time-resolved WAXS experiment for $D_{77}^{32}C_{23}^{10}$ (a PPDX-*b*-PCL diblock copolymer) performed to monitor changes in the crystal structure on quenching from the melt from 115 to 30 °C. Figure 7 shows results from analysis of the integrated WAXS peak areas, as well as the relative degree of crystallinity determined from the WAXS data. A clear two-stage crystallization process occurs. First the peaks at $2\theta = 22°$ and 23.8° grow as a result of

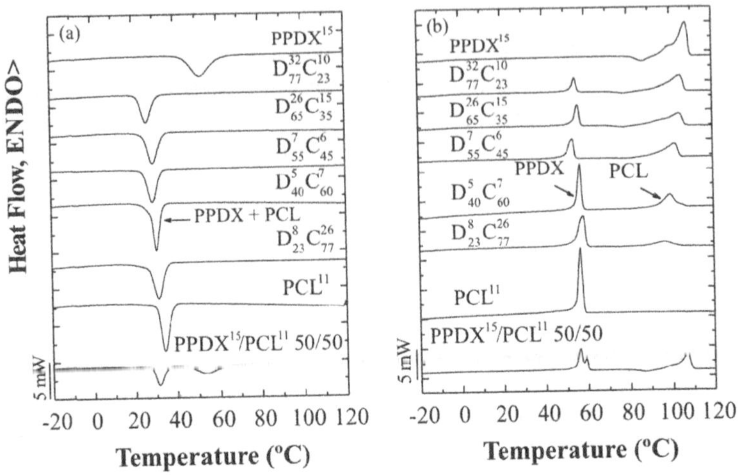

Fig. 6 **a** DSC cooling scans ($10\,°\mathrm{C\,min^{-1}}$) for poly($\varepsilon$-caprolactone) (*PCL*) and poly(*p*-dioxanone) (*PPDX*) homopolymers, diblock copolymers and a 50/50 blend. **b** Subsequent heating scans ($10\,°\mathrm{C\,min^{-1}}$). (From [103]. Reproduced with permission of the Royal Society of Chemistry)

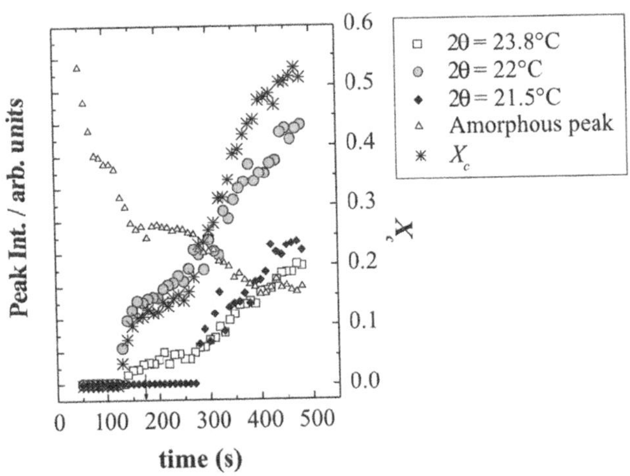

Fig. 7 Relative integrated intensities under the peaks (*left scale*) and relative degree of crystallinity (*right scale*) obtained from wide-angle X-ray scattering data for $D_{77}^{32}C_{23}^{10}$ quenched from $T = 115$ to $30\,°\mathrm{C}$. The cooling rate was $30\,°\mathrm{C\,min^{-1}}$. The *arrow* indicates the nominal point at which the final temperature was reached. For convenience, the amorphous peak intensity has been divided by 5. (Reprinted with permission from [107]. Copyright 2003 American Chemical Society)

PPDX crystallization, reaching a plateau prior to a further increase in integrated area upon PCL crystallization. The PCL crystallization is clearly seen in the growth of the $2\theta = 21.5°$ peak, which can be ascribed to an exclusive crystalline PCL reflection after an induction time [107]. These results support DSC evidence obtained during isothermal crystallization (see later) that points to a nucleating effect of PPDX on PCL, since they have helped to establish that even under isothermal conditions the PPDX block crystallizes first, closely followed by the PCL block.

Going back to Fig. 6, during cooling at 10 °C min^{-1} the PPDX block needs greater supercooling to start its crystallization. Once the crystallization of the PPDX block starts, it is quickly followed by the crystallization of the PCL block, which can be nucleated by the PPDX block. As a result, the crystallization of both blocks occurs in the same temperature range and in a quick succession leading to an overlap of the DSC exothermic signals during cooling. This overlap cannot be resolved by the use of slower cooling rates, at least down to 1 °C min^{-1}. This may be related to the fact that the PCL block only crystallizes after the PPDX block starts to crystallize; this was clearly demonstrated by previous WAXS results [107]. One way to separate completely the crystallization signals is by self-nucleating the PPDX block, since this shifts the crystallization temperature of the PPDX to much higher temperatures [102, 103].

In a related double-crystalline PE-*b*-poly(L-lactide), PE-*b*-PLLA diblock copolymer (with an approximate 50/50 weight ratio), similar behavior was observed [103]. When the diblock copolymer is cooled from the melt at 10 °C min^{-1} only one crystallization exotherm is observed, indicating that the L-lactide block is also affected by the molten PE block that is covalently bonded to it [103]. However in this case, the crystallization exotherms can be separated by reducing the cooling rate to 2 °C min^{-1} [106].

Figure 8a shows experimentally determined spherulitic growth rates as a function of crystallization temperature for $D_{77}{}^{32}C_{23}{}^{10}$; they were performed at temperatures where the PCL block is molten. The difference in growth rates is very pronounced between the homopolymer and the PPDX block within the diblock copolymer and amounts to an order of magnitude at certain crystallization temperatures. The effect of the molten PCL chain covalently bonded to PPDX is unexpectedly large. This is reflected in the energy barrier for secondary nucleation as calculated by fitting the Lauritzen and Hoffman theory to the data presented in Fig. 8a by using the typical plots shown in Fig. 8b. From Fig. 8b, values of the slope (i.e., K_g) represent an energy term that quantifies the value of the energetic restrictions for the crystallization of the chains. Figure 8b shows the values obtained for K_g for PPDX^{15} and the PPDX block within $D_{77}{}^{32}C_{23}{}^{10}$. It can be observed that the K_g value for the PPDX block within the copolymer is more than twice that of neat PPDX^{15}. As expected from the K_g values, both the fold surface free energy and the work for chain folding will be much higher for the PPDX block within $D_{77}{}^{32}C_{23}{}^{10}$ than for the PPDX^{15} homopolymer [103]. The application of the Lauritzen and Hoffman theory to overall crystallization rate data obtained by DSC was

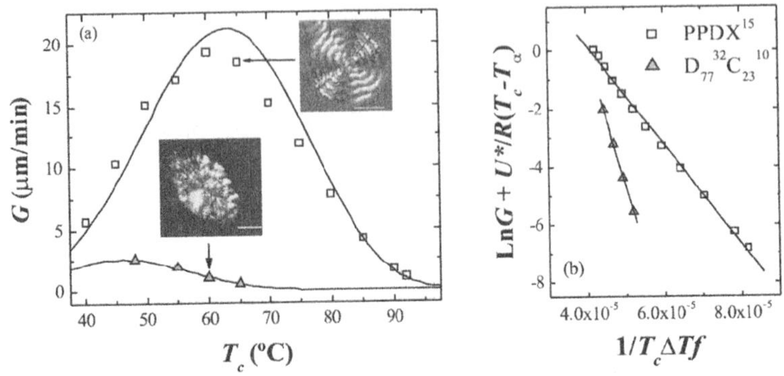

Fig. 8 **a** Spherulitic growth rates for PPDX and the PPDX block within $D_{77}{}^{32}C_{23}{}^{10}$ diblock copolymer. *Solid lines* are fits to Lauritzen and Hoffman theory. **b** Lauritzen and Hoffman kinetics theory plot for PPDX ($K_g^G = 17.2 \times 10^4\ K^2$) and the PPDX block within $D_{77}{}^{32}C_{23}{}^{10}$ diblock copolymer ($K_g^G = 46 \times 10^4\ K^2$). (From [103]. Reproduced with permission of the Royal Society of Chemistry)

also highly successful for PPDX and the diblock copolymers. This was corroborated by a novel procedure that consisted of applying the Lauritzen and Hoffman theory to isothermal crystallization data from DSC measurements of previously self-nucleated PPDX and obtaining a perfect match with the energetic parameters determined from spherulitic growth rate data. The use of self-nucleation allowed the determination of the overall crystallization kinetics of the PPDX block within two copolymers (with 65 and 77% PPDX) which was otherwise too slow for detection. The results showed that the energetic restrictions for the crystallization of the PPDX block at high temperatures increase as the PCL content in the copolymer increases. The effect of the molten PCL block can also affect the spherulitic superstructure as seen in Fig. 8a, where polarized optical micrographs are presented. For equivalent high temperatures, in the copolymer case, the spherulites are deformed, exhibiting an ellipsoidal shape and no regular banding compared with PPDX homopolymer [107].

The dramatic depression of the growth rate in the PPDX block within the $D_{77}{}^{32}C_{23}{}^{10}$ diblock copolymer compared with the growth rate of an equivalent molecular weight homopolymer is the determining factor in the coincident crystallization behavior encountered upon cooling the copolymers from the melt in Fig. 6. Recently, Ueda et al. [26] examined the crystallization kinetics of a flow-oriented PE-*b*-(atactic polypropylene), PE-*b*-aPP, with a PE volume fraction 0.48. In this case, the crystallization of the PE block occurs with a molten aPP block covalently bonded to it. The components are reported to be in the strong segregation limit and therefore are phase-separated well before crystallization. They found that the crystallization kinetics (as determined by small-angle light scattering) was substantially retarded compared with that of an equivalent molecular weight PE homopolymer. They attributed the effect to a mobility reduction for the chains close to the interfacial region and to the presence of the noncrystallizable aPP chains close to the growth face which could obstruct the growth process. These results are similar to those obtained for PPDX-*b*-PCL diblock copolymers [107] for the crystallization of the PPDX block at higher temperatures, when the PCL component is molten. Similar results were also obtained by Shiomi et al. [113] since they detected a reduction in crystallization rate when they compared a PEO homopolymer with the PEO block within PEO-*b*-PB diblock copolymers. In this case, the crystallization of the PEO block was carried out with a covalently bonded rubbery block and similar energetic restrictions for its crystallization decrease its crystallization rate compared with that of PEO homopolymer.

In order to study the overall crystallization kinetics of the PCL block within PPDX-*b*-PCL diblock copolymers, Müller et al. [103] first crystallized the PPDX block until saturation by performing a special thermal procedure (it consisted of first cooling from the melt as in Fig. 6 to allow both blocks to crystallize, then the sample was heated to 62 °C and annealed at that temperature for 70 min, a temperature at which the PCL is molten, before quenching to a T_c where the PCL block isothermal crystallization was followed by DSC). With the use of such a procedure the overall isothermal crystallization of only the PCL block was determined in the diblock copolymers where the PPDX block was already crystallized.

The overall crystallization rate, expressed as the inverse of the experimental crystallization half-time, for PCL^{11} and for the PCL block of the diblock copolymers employed are shown in Fig. 9, where the nucleation effect of the previously crystallized PPDX block on the PCL block is confirmed. The overall crystallization rate of the PCL block in the case of $D_{23}{}^{8}C_{77}{}^{27}$ and $D_{40}{}^{5}C_{60}{}^{7}$ is higher than that of PCL^{11}. This effect was reported in a previous work when the PPDX content in the copolymer was higher than 53% [102]. In cases where the PPDX content in the diblock copolymer was higher than or equal to 53%, the nucleation effect of the PPDX block seems to be overtaken by topological restrictions. Figure 9 shows that the PCL blocks within $D_{77}{}^{32}C_{23}{}^{10}$ and $D_{65}{}^{26}C_{35}{}^{15}$ crystallize at a slower overall rate than neat PCL. In these cases, the PPDX matrix has already crystallized in spherulites and the 23 or 35% PCL must crystallize within them. Even though the nucleating effect of PPDX on PCL prevents any homogeneous nucleation, increasing amounts of PPDX within the PPDX-*b*-PCL diblock copolymers hinder crystallization of the PCL block as already indicated by an overall reduction in kinetics (Fig. 9). Another sign of the topological restrictions encountered by the PCL block was reported since the Avrami index was found to decrease as the PCL content in the copolymer decreased [102, 103]; however, its value ranged from 3.5 to 2.1. Such a decrease in the Avrami index can be interpreted as a decrease in the dimensionality of growth of the PCL block superstructures as more previously crystallized PPDX is present in the sample.

The results of Müller et al. [103] on PPDX-*b*-PCL diblock copolymers differ from those obtained previously by Bogdanov et al. [105] when they studied by DSC the crystallization kinetics of 80/20 PCL-*b*-PEO diblock copolymers. In their case, the PCL block crystallized first from a homogeneous melt and the Avrami parameters *K* and *n* were found to be similar to the kinetics parameters of the isothermal crystallization of a corresponding PCL homopolymer. Significant crystallization retardation was found for the PEO block that crystallized second. The retardation was attributed to the "mutual influence between the PEO constituent and the PCL crystal phase which fixes (hardened) the total copolymer structure" [105]. In the PPDX-*b*-PCL diblock copolymer case, conversely, when the PPDX block crystallizes it does so at a slower rate than a comparable PPDX homopolymer. The crystallization of the PCL block, on the other hand, strongly depends on the composition of the diblock copolymer as shown in Fig. 9. However, for $D_{77}^{32}C_{23}^{10}$, the overall kinetics is retarded, which could be regarded as similar to the retardation experienced by the PEO block in the 80/20 PCL-*b*-PEO diblock copolymer studied by Bogdanov et al. [105]; the Avrami index in both cases was also of the order of 2.

Another double-crystalline system that has recently been studied is that of PLLA-*b*-PCL diblock copolymers [110]. Kim et al. synthesized rather broad molecular weight distribution PLLA-*b*-PCL diblock copolymers with polydispersities ranging from 1.3 to 1.6. They reported that a diblock copolymer with a M_n value of 77 000 and a PCL weight fraction of 0.32 was phase-separated

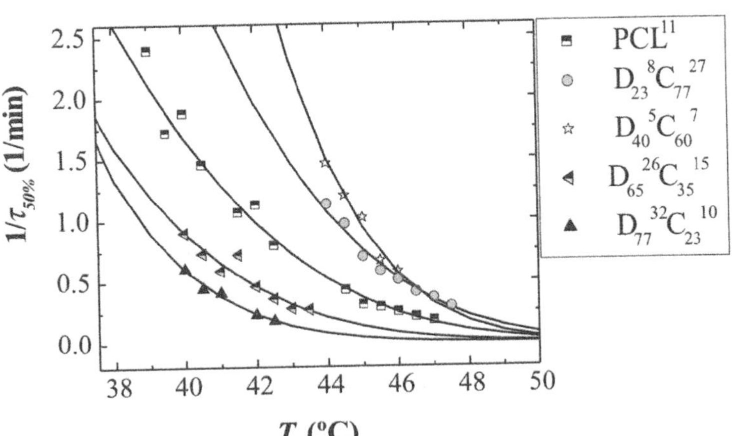

Fig. 9 Inverse of the crystallization half-time as a function of isothermal crystallization temperature for PCL^{11} homopolymer and for the PCL block of the indicated copolymers. All experiments were performed after the PPDX block had been previously crystallized until saturation. Solid lines are fits to the Lauritzen and Hoffman theory. (From [103]. Reproduced with permission of the Royal Society of Chemistry)

in the melt at 220 °C according to rheological measurements; however, SAXS failed to show any signs of phase segregation. When they employed another sample with a lower molecular weight ($M_n = 19\,000$), the behavior was that of a homogeneous melt above the PLLA block melting temperature. More recently, Ho et al. [108] prepared a 33 900 g mol^{-1} sample of PLLA-*b*-PCL copolymer with a 1.2 polydispersity and a volume fraction of PCL of 0.28. They concluded that the sample was melt-mixed and their work showed that it was possible to use epitaxial crystallization to obtain large, well-oriented MDs of PLLA-*b*-PCL by employing several crystalline substrates, including benzoic acid and hexamethylbenzene. When an amorphous substrate was employed the sample crystallized in spherulites as expected for a homogeneous melt copolymer.

5 ABC Triblock Copolymers with One Crystallizable Block

Contributions to the subject of ABC triblock copolymer crystallization are listed in Table 1, where some characteristics of the triblock copolymers involved are reported with the corresponding references.

5.1 Crystallization and Melting Behavior

Even though the first report about the synthesis of crystallizable ABC triblock copolymers was published in 1978 for PS-*b*-PB-*b*-PCL copolymers [114], in that work only a preliminary study of the tensile properties was performed, without considering the crystallizability of the materials. It was only 20 years later, when the preparation of these materials was reconsidered and optimized, that triblock copolymers with relatively narrow molecular weight distributions were obtained [115], a requisite which is indispensable for the generation of well-defined morphologies. To illustrate the complexity and richness of semicrystalline ABC triblock copolymers, PS-*b*-PB-*b*-PCL triblock copolymers have been chosen. These copolymers have been prepared with a wide composition range (with PCL contents from 11 to 77%) and they have been compared with PS-*b*-PCL and PB-*b*-PCL diblock copolymers [29, 98, 115–118].

Calorimetric measurements and morphological observations showed that PS-*b*-PCL, PB-*b*-PCL and PS-*b*-PB-*b*-PCL copolymers exhibit microphase separation and crystallization if the molecular weight is high enough. Only in PS-*b*-PCL diblock copolymers, a shift of the PS glass transition to lower temperatures has been observed. In PS-*b*-PB-*b*-PCL, the crystallizable block (i.e., PCL) is covalently linked to a rubbery block and it has a free end. For these reasons there is no significant reduction in the melting temperature and crystallinity degree of the PCL block when its content is higher than about 50%, when compared with an analogous homopolymer. Even, when the PCL content is between 16 and 40% a melting point reduction of only 4–6 °C has been observed [29, 116, 118]. Similar results were also found by Schmalz et al. for PB-*b*-poly(isoprene)-*b*-PEO (PB-*b*-PI-*b*-PEO), who investigated in detail block copolymers with a minority of the crystallizable component (PEO). In this type of block copolymer the PB and PI blocks form a mixed phase with a glass transition of about – 70 °C. Consequently, they might be considered as diblock copolymers consisting of a PEO phase and a mixed PB/PI phase, in which the PEO block is strongly segregated from the other phase [119]. In addition, Schmalz et al. found that T_m increased with increasing PEO content. Thus, they observed variations from $T_m = 60.0$ °C for $B_{11}I_{70}EO_{19}$ to $T_m = 65.9$ °C for $B_{19}I_{39}EO_{42}$, where the PEO blocks build spherical and lamellar MD, respectively.

Bailey et al. [120] and Epps et al. [121] characterized the behavior of other block copolymer systems with PEO, PI-*b*-PS-*b*-PEO and PS-*b*-PI-*b*-PEO, and they studied the effect of lithium perchlorate addition to the copolymers. They maintained constant the PS to PI ratio and varied the PEO volumetric fraction from 0 to 33%. The comparison of both types of triblock copolymers showed that the melting temperature of the PEO block is lower when it is directly attached to PS. This was attributed to the glassy character of the PS block. On the other hand, when it is linked to PI, which is rubbery at the crystallization temperature of the PEO block, the melting temperatures for PEO are higher. It should be mentioned, however, that the crystallinity degrees in both series of copolymers (PI-*b*-PS-*b*-PEO and PS-*b*-PI-*b*-PEO) are of the same order of magnitude, independent of the neighboring block. Then, the number of crystals is probably higher in those copolymers with a PS block linked to PEO, but their lamellar sizes are smaller as indicated by their lower T_m values. These are very illustrative examples of the influence of the block copolymer topology (Fig. 10). The reduction of T_m is more marked when the PEO nanophases are smaller, like in the results found by Schmalz et al. for PB-*b*-PI-*b*-PEO triblock copolymers. They also reported cold crystallization for most of the copolymers as a result of PEO confinement.

Even though the thermal behavior upon heating after a previous controlled cooling in the DSC of the crystallizable block does not exhibit significant changes when the neighboring block is rubbery, the crystallization is influenced by the content of the crystallizable block and, consequently, by the block copolymer morphology. If the crystallizable block is confined within MDs, crystallization can take place at much larger supercoolings than usual. This has been attributed to the occurrence of fractionated crystallization and/or homogeneous nucleation [29]. In Sects. 2 and 3 the phenomenology of the fractioned crystallization and homogeneous nucleation for confined homopolymers and diblock copolymer components was discussed in great detail. In this section we will present some examples of such a phenomenon in

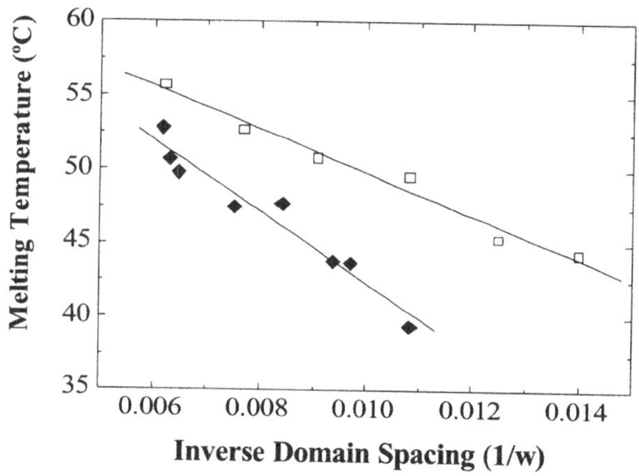

Fig. 10 PEO melting temperatures vs. inverse PEO domain size as a function of block architecture (*squares*, PS-*b*-PI-*b*-PEO; *diamonds* PI-*b*-PS-*b*-PEO). *PI* poly(isoprene); *w* is given in angstroms. (Reprinted with permission from [121]. Copyright 2003 American Chemical Society)

the particular case of ABC triblock copolymers. To our knowledge, Balsamo et al. [116] were the first to report fractionated crystallization in the case of ABC triblock copolymers. They detected the appearance of two crystallization exotherms at about 30 and – 45 °C in $S_{57}B_{27}C_{16}$ during a cooling scan performed in the differential scanning calorimeter. The results were explained by considering that the higher temperature crystallization exotherm was produced by heterogeneous nucleation and the lower temperature crystallization exotherm by homogeneous nucleation, owing to its close value to the PCL glass transition [116]. Recently, Müller et al. [29] and Schmalz et al. [101, 119] published extended works, where they investigated a wide selection of AB and ABC triblock copolymers containing one or two components that exhibited fractionated crystallization. This wide selection of copolymers allowed confinement of crystallizable blocks to be studied in very different types of MD morphologies, like cylinders, spheres, undulated layers, rings and lamellae.

As mentioned in Sect. 3, for PEO it has been found that the crystallization temperature is often a function of the MD volume. The examples quoted in Sect. 3 referred to PEO dispersed in droplets or to PEO that was a component within diblock copolymers. For other block copolymer components like PCL the variation in T_c encountered upon MD size increase is not as pronounced. Nojima et al. [22] found that the variation of T_c for PB-*b*-PCL block copolymers with spherical PCL MDs of increasing sizes, ranging from 10.3 to 17.4 nm, was of about 5 °C for crystallization at very large supercoolings (T_c fluctuated between – 50 and – 45 °C approximately). For ABC triblock copolymers, Müller et al. [29], Schmalz et al. [101, 119] and Balsamo et al. [118] found, by studying copolymers with minority components of PEO or PCL blocks linked to a rubbery block, that the T_c associated with fractionated crystallization does not show any strong dependence of the morphological features of the MD and of the molecular weight of the crystallizable block (Table 4).

As an example, it can be observed in Table 4 that $S_{62}B_{27}C_{11}$ and $S_{50}B_{28}C_{22}$, which form core–shell cylinders with PCL cores and have PCL molecular masses of 6820 and 21 340 g mol^{-1}, have crystallization exotherms located at – 45.1 and – 45.3 °C, respectively. On the other hand, $S_{20}B_{41}C_{39}$, which exhibits a lamellar–lamellar morphology and have an even higher PCL molecular mass, 51 480 g mol^{-1},also shows a crystallization exotherm at similar supercoolings (– 43.7 °C). In the same way, $B_{11}I_{70}EO_{19}$, $B_{17}I_{57}EO_{26}$ and $B_{24}I_{56}EO_{20}$, which have PEO molecular masses of 22 800, 13 400 and 33 800 g mol^{-1}, have at least a crystallization exotherm at temperatures as low as – 23, – 25 and – 21 °C, respectively. These latter data are not in agreement with the results of Xu et al. [96] for PB-*b*-PEO diblock copolymers with relatively large PEO blocks (Sect. 3).

The data of Table 4 clearly indicate that the PEO block within PB-*b*-PI-*b*-PEO triblock copolymers exhibits fractionated crystallization when it constitutes a minority component. $B_{11}I_{70}EO_{19}$ and $B_{24}I_{56}EO_{20}$ show only one crystallization exotherm, whereas $B_{17}I_{57}EO_{26}$ has two crystallization exotherms.

Table 4 Peak crystallization temperatures determined by DSC during cooling from the melt of ABC triblock copolymers and selected homopolymers [29, 101, 118, 119]

System	Morph*a	T_c (I) (°C)	T_c (II) (°C)
PEO^{1}		21.0	
PEO^{100}		43.0	
PCL^{32}		33.5	
$B_{11}I_{70}EO_{19}{}^{120}$	PEO Sph		– 23.0
$B_{24}I_{56}EO_{20}{}^{67}$	–		– 25.0
$B_{17}I_{57}EO_{26}{}^{130}$	PEO Cyl	15.0	– 21.0
$B_{19}I_{39}EO_{42}{}^{135}$	PEO Lam	37.5 (19.8)	– 25.0
$S_{62}B_{27}C_{11}{}^{62}$	PCL core in c_ic	20.4	– 45.1
$S_{37}B_{52}C_{11}{}^{96}$	PCL core in c_ic	6.4	– 48.4
$S_{57}B_{27}C_{16}{}^{137}$	PCL core in c_ic	30.0	– 45.3
$S_{26}B_{36}C_{38}{}^{110}$	PCL Lamellae in ll	36.0	– 40.8
$S_{20}B_{41}C_{39}{}^{132}$	PCL Lamellae in ll	16.7	– 43.7
$S_{51}B_{09}C_{40}{}^{98}$	PCL Lamellae in lc	22.7	–

*a Lam(Lamellar); Cyl(Cylinders); Sph(Spheres) c_ic: core shell cylinders, ll: lamellar-lamellar, lc: lamellar-cylindrical.

On the other hand, $B_{19}I_{39}EO_{42}$ exhibits peak crystallization temperatures as high as 37 and 20 °C, which are close to the values observed in PEO homopolymer, but it also has a small crystallization exotherm at – 25 °C. Since PS-*b*-PEO diblock copolymers with minority PEO content crystallize at larger supercoolings when they are homogeneously nucleated (at about – 40 °C) [29], the crystallization exotherms located at around – 25 °C were attributed either to a weak nucleating influence of the interphase or to the existence of other types of heterogeneities that are only active at such large supercoolings (since no dependence on MD size seems apparent in the case of PEO within ABC triblock copolymers). The presence of additional exotherms at higher temperatures is indicative of the presence of MD interconnections. The same explanation applies for PS-*b*-PB-*b*-PCL triblock copolymers, which in addition to the exotherm present at around – 45 °C, also show crystallization exotherms between 6 and 30 °C.

Chen et al. [87] have argued that the nucleation events they observed in confined MD structures within PB-*b*-PEO diblock copolymers or PB/PB-*b*-PEO blends must be homogeneous since impurities cannot be contained within the nanoscopic MD present. In their case, the nucleation was most probably homogeneous in view of the large supercoolings obtained but not because impurities are necessarily excluded from the MD. Müller et al. [29] demonstrated that the occurrence of heterogeneous nucleation within confined nanoscopic MD structures is certainly possible through the example provided in the following.

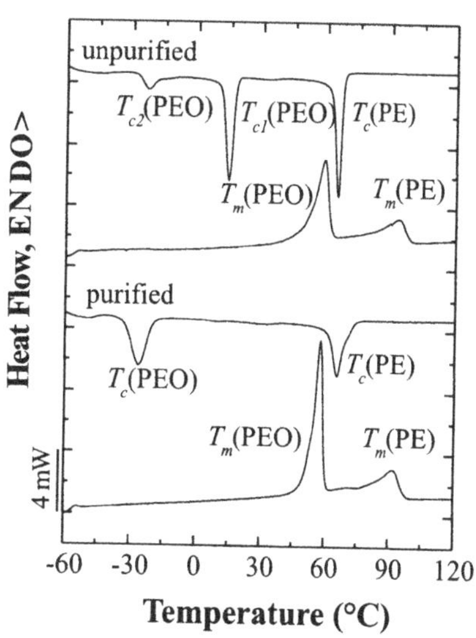

Fig. 11 Cooling and heating DSC scans (10 °C min^{-1}) of original and purified $E_{24}EP_{57}EO_{19}$. (Reprinted with permission from [29]. Copyright 2002 American Chemical Society)

Figure 11 shows the fractionated crystallization of the PEO block within $E_{24}EP_{57}EO_{19}$, where E is PE and EP is a poly(ethylene-*co*-propylene) copolymer. This triblock copolymer was obtained after hydrogenation of $B_{24}I_{56}EO_{20}$ using a Wilkinson catalyst. The cooling scan of the original $E_{24}EP_{57}EO_{19}$ shows a crystallization exotherm for the PE block at 69 °C. As the temperature is decreased, a large part of the PEO crystallizes just below 20 °C, while a small fraction of the PEO can only crystallize at much lower temperatures (– 20 °C). In this case, the PEO block is probably confined into cylinders. Then, the crystallization observed at around 20 °C must be due to the presence of heterogeneities in the polymer. In order to demonstrate this hypothesis, the same sample was subjected to a purification process to extract the Wilkinson catalyst. The result is also shown in Fig. 11. The exotherm that was present at around room temperature has completely disappeared, and all of the PEO now crystallizes at temperatures lower than – 27 °C. In fact, two low crystallization exotherms for the PEO block were identified: one peak at around – 27 °C and a small exotherm at – 47 °C. Probably, the first exotherm corresponds to heterogeneous nucleation from a weakly nucleating heterogeneity or surface nucleation and the lowest crystallization exotherm originated after homogeneous nucleation of the PEO block. Similar results were obtained by Balsamo et al. [116] after purification of $S_{57}B_{27}C_{16}$. This subject will be considered in detail in the discussion of the triblock copolymers with two crystallizable blocks, where self-nucleation experiments were done to support this explanation.

Another aspect that has been investigated in semicrystalline ABC triblock copolymers is the effect of different annealing conditions at high temperatures (to induce a higher order in the morphology) maintaining invariable the crystallization conditions. This kind of study has been carried out in PS-*b*-PB-*b*-PCL [118, 122]. Although the influence was stronger in triblock copolymers with relatively high PCL content, it was also observed that the increase of the annealing time at 150 °C in a copolymer with a small PCL content led to the definition of different lamellar thicknesses within the PCL MD. Confining the semicrystalline component into lamellae, in contrast to confining it into cylinders, leads to higher "conflict" between crystallization and microphase separation. Thus, when the PCL block is less confined there is a clear conflict between the thermodynamic force to form, for example, spherulitic structures and the thermodynamic force to maintain the microphase-separated morphology formed in the melt. It seems to be that lamellar thickening is less restricted when formation of spherulites does not occur. This aspect will be considered again in Sect. 5.2.

Even though the nonisothermal crystallization leads to just small changes in the subsequent melting behavior of different types of triblock copolymers, isothermal experiments employed to calculate the equilibrium melting temperature, T_m^o, have shown that this parameter can exhibit significant changes depending on composition. It has been reported that in PS-*b*-PB-*b*-PCL triblock copolymers there is a marked decrease of T_m^o with an increase of the weight ratio $w_{PB} : w_{PCL}$ [116]. This indicates a thinning of the crystalline lamellae as a consequence of the connection of the PCL block to PB. Only for very short middle blocks was an influence of the PS block on the equilibrium melting point observed. The change of the lamellar thickness, and hence of T_m^o, can be understood by considering two opposing effects. The system tends to form an extended PCL block with minimal folding to lower the free energy of the crystalline region. The amorphous chains, on the other hand, prefer to maintain a random conformation [123, 124].

The isothermal crystallization kinetics can be analyzed employing the Avrami equation (Eq. 7). In Sect. 3 the direct relationship that exists between the results given by the Avrami analysis and the confinement degree of a crystallizable component in diblock copolymers was discussed. Balsamo et al. [118] studied the crystallization kinetics of ABC triblock copolymers with different morphologies. Avrami exponents, n, higher than 1 were obtained in copolymers that form PCL lamellar MDs. This revealed along with the nonisothermal behavior that the nucleation step is taking place heterogeneously and that the MD should be interconnected as was thoroughly discussed for diblock copolymers in Sect. 3. Therefore, it can be concluded that for block copolymers in general (diblocks and triblocks), the Avrami exponent effectively reflects the dimensionality of crystal growth when crystallization proceeds from heterogeneous nucleation.

5.2 Morphology

A controversy has arisen as to whether the observations by POM and those by transmission electron microscopy reflect the same morphological features or not. In fact, Kim et al. [125] demonstrated that the same block copolymer can exhibit different morphologies depending on sample thickness, this being a possible reason for the sometimes contradictory results found in several works. Nevertheless, before this aspect can be properly treated in this section, we present a review of the morphological investigations carried out in semicrystalline ABC triblock copolymers at a nanoscopic scale.

An example of the incorporation of an external component on the crystallization behavior of triblock copolymers was given by Schmalz et al. [126]. They obtained PS-*b*-poly(ethylene-*co*-propylene)-*b*-PE (PS-*b*-PEP-*b*-PE) triblock copolymers from the hydrogenation of PS-*b*-PI-*b*-PB. PS-*b*-PEP-*b*-PE triblock copolymers have the peculiarity that PEP and PE have an interaction parameter of 0.007 at 120 °C; therefore, they form a homogeneous melt, which is segregated from the PS block and can be considered as an intermediate case between diblock and triblock copolymers. The crystallization of the PE block occurs at about 60 °C and the authors evaluated the influence of the incorporation of a solvent during the crystallization and segregation processes under two conditions. When no solvent was present in the system, PE crystallization took place from the melt, the PS vitrified within the PEP matrix and finally the PE crystallized within distorted cylinders. When crystallization occurred from a toluene solution, which is a good solvent for PS, PE crystallized before PS vitrification, then PE cylinders were formed within a PEP matrix and finally PS formed distorted cylinders between the PE crystals. As can be seen, similar to the effect found in some diblock copolymers, the incorporation of a selective solvent leads to changes in the MD structure. In this case, toluene delays the PS vitrification in comparison with PE crystallization (Fig. 12).

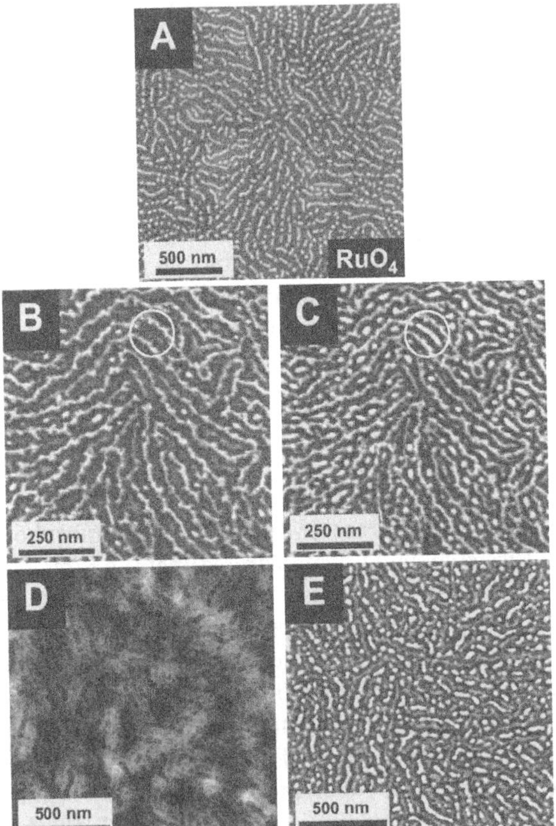

Fig. 12 **a** Scanning electron microscpy image of a dip-coated film of $S_{13}EP_{57}E\_30^{112}$ onto a silicon wafer. **b** and **c** Tapping-mode atomic force microscopy (*AFM*) phase-contrast images of a thin film of $S_{13}EP_{57}E\_30^{112}$ dip-coated onto a silicon wafer: **b** dry film; **c** same spot of the film after 1-min exposure to toluene vapor, visualizing the PS cylinders; $z = 20°$. Tapping-mode AFM height (**d** $z = 15$ nm) and phase (**e** $z = 50°$) image of a thin film of $S_{14}EP_{64}E\_22^{112}$ spin-coated from a 5 wt % solution in toluene onto a silicon wafer. (Reprinted with permission from [126]. Copyright 2001 American Chemical Society)

Even though the works of Bailey et al. [120, 127] and Epps et al. [121] present very comprehensive studies about the variation of the morphology as a function of composition and topology, they did not study in detail the relationship "microphase separation–crystallization". Several contributions have been made in this area, when the triblock copolymers have only one crystallizable block [101, 118, 119, 122, 126, 128–130]. Some relevant aspects of these references have already been mentioned.

For semicrystalline–glassy block copolymers it is well known that there is a trend to preserve the self-assembled structure established in the melt as long as strong segregation prevails. However, two intriguing aspects emerged from the investigations: how do the crystalline lamellae coexist with the amorphous glassy MD in copolymers with a majority of the semicrystalline component? and how does the system accommodate the density change upon crystallization in copolymers with confined semicrystalline components? Balsamo et al. [129] published a first contribution where this aspect was discussed for PS-*b*-PB-*b*-PCL triblock copolymers. Even though the PCL crystallization process takes place at temperatures well below the T_g of the PS block, copolymers with high PCL content experience a deformation of the glassy MD upon crystallization to generate, for example, ellipsoidal core–shell cylindrical microphases of PS and PB embedded in a PCL matrix.

In the case in which the semicrystalline component is confined, Register and Loo [44] have argued that three possibilities exist to accommodate the density change that occurs upon crystallization: existence of individual crystallized MDs under significant hydrostatic tension, cavitation within the crystallizing domains and macroscopic contraction to accommodate the volume change. From these, Loo et al. [42] demonstrated in PE-*b*-PVCH copolymers that the third possibility is correct. Though the PVCH matrix is vitreous throughout crystallization and melting, it can deform as necessary so that PE domains neither cavitate nor exist under great overall tension. This is in agreement with literature results that have shown evidence of deformation of confined core–shell cylinders as can be observed in Fig. 13 [119, 128]. In contrast to the core–shell cylinders observed for amorphous triblock copolymers of similar composition, the cylinders in Fig. 13 have an unusual polygonal shape. This has been attributed to the deformation of circular cylinders during the crystallization process within inner cylinders. In PE-*b*-PEP-*b*-PEO block copolymers this could be expected since the matrix is rubbery (PEP); however, in the case of the PS-*b*-PB-*b*-PCL triblock copolymer, the matrix is glassy (PS). This indicates, in agreement with the results of Loo et al. [42], that deformation is necessary in order to accommodate the density change that occurs upon crystallization.

The study of the microphase morphology in several types of crystallizable triblock copolymers has shown that when the triblock copolymers have compositions from which the crystallizable component is able to build lamellae, there is a tremendous influence on the mesophase structure. The MDs do

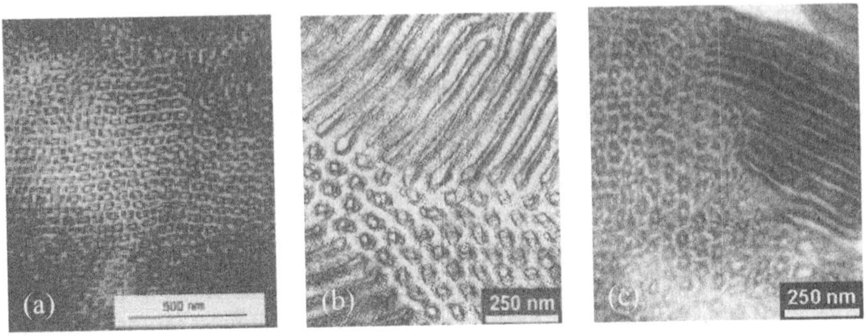

Fig. 13 Transmission electron microscopy (*TEM*) micrographs of **a** $S_{57}B_{27}C_{16}$: the *white*, *gray* and *black areas* correspond to PCL, PS and poly(butadiene) (*PB*), respectively, after staining with OsO_4 [128]. **b** $E_{19}EP_{40}EO_{41}$: only the interphase between PEO cylinders and the Poly(ethylene-*ran*-propylene) matrix gets preferentially stained with RuO_4 [119]. **c** TEM morphology for $E_{19}EP_{40}EO\_41^{138}$ hydrogenated with *p*-toluenesulfonyl hidrazide and stained with RuO_4 [119]. (Reprinted with permission from [119, 128]. Copyright 1999 and 2002 American Chemical Society)

not show the same degree of long-range order as in amorphous ABC triblock copolymers; moreover, they can be deformed as already mentioned [125, 130]. In addition, a change of the crystallization conditions can induce a morphological transition. For example, copolymers that form a lamellar–lamellar morphology, when crystallized at a specific T_c, can experience a morphological transition, when the crystallization temperature is reduced, to a lamellar–cylindrical morphology owing to a rupture of the amorphous lamellae when the crystalline lamellar thickness is reduced. This occurs in order to avoid a thermodynamically unfavorable stretching of the amorphous block [130]. The microphase morphology can also be modified by the annealing conditions at high temperature. Thus, in $S_{51}B_{09}C_{40}$ a morphological transition occurs from a lamellar–cylindrical to a cylindrical–ring morphology, as a means to reduce contacts between the PCL and PB blocks as can be appreciated in Fig. 14 [122].

We have already mentioned that depending on composition, semicrystalline triblock copolymers can show some “conflict” between microphase separation and superstructure formation. In fact, one of the controversial aspects is the question whether block copolymers can or cannot exhibit spherulites. This is a relevant question because spherulitic structures greatly affect the ultimate mechanical properties, and the boundaries between adjacent spherulites are often weak points in mechanical performance. Kim et al. [125] studied the competition between crystallization within microphase-separated regions and reorganization into supermolecular spherulites in semicrystalline PS-*b*-PB-*b*-PCL triblock copolymers. These authors found that the formation of spherulites is strongly affected by the thickness of the specimen in such a way that thin films crystallize into

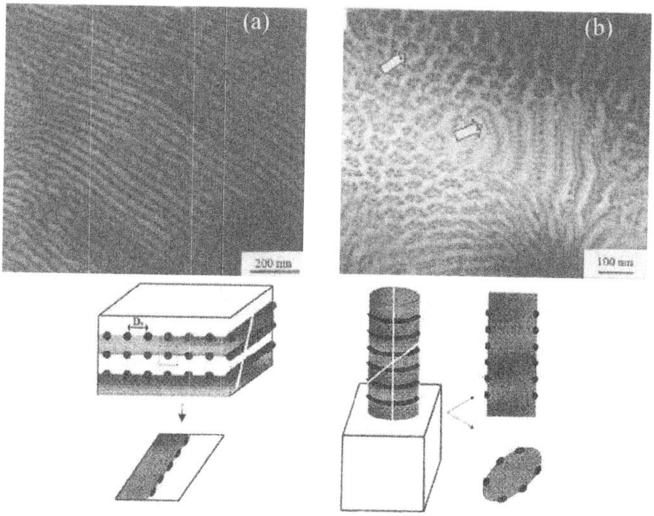

Fig. 14 TEM images of $S_{51}B_{09}C_{40}$ after thermal treatment at 150 °C for **a** 5 h and **b** 15 hours. (Reprinted with permission from [122]. Copyright 2003 American Chemical Society)

spherulites more easily than bulk films. This effect can be appreciated in Fig. 15, where the microphase morphology and superstructure can be compared for bulk (Fig. 15a, b) and thin (Fig. 15c, d) films of $S_{35}B_{15}C_{50}$ without and after thermal treatment at 130 °C.

In Fig. 15 it can be observed that a microphase-separated lamellar–cylindrical morphology is formed in the as-cast and thermally treated triblock copolymer bulk films (Fig. 15a, b). In the inset of Fig. 15a, a speckle pattern of birefringence in POM confirms the crystallization of PCL within the microphase-separated structures, but if spherulites or precursors are present, they must be larger than 2 μm. Although in the as-cast bulk sample evidence of morphological deformation from the local crystallization of the PCL is not observed, except for a slight waviness of the lamellae, the thermally treated bulk sample shows regions with somewhat better long-range lamellar–cylindrical order. Presumably, reorganization into spherulites cannot occur because the PS lamellar phase is glassy well above the crystallization temperature of PCL. In contrast, thin films of the same sample exhibit a morphological transition. The as-cast thin films show a similar lamellar–cylindrical morphology with a speckle pattern in POM as described for the bulk films (Fig. 15c). However, upon thermally treating the sample at 140 °C, the microphase morphology is no longer lamellar–cylindrical, but is instead replaced by lamellar structures, which are located within spherulites as can be demonstrated by POM (Fig. 15d). Although the reason for this is not completely clear, the authors considered that the explanation for the different

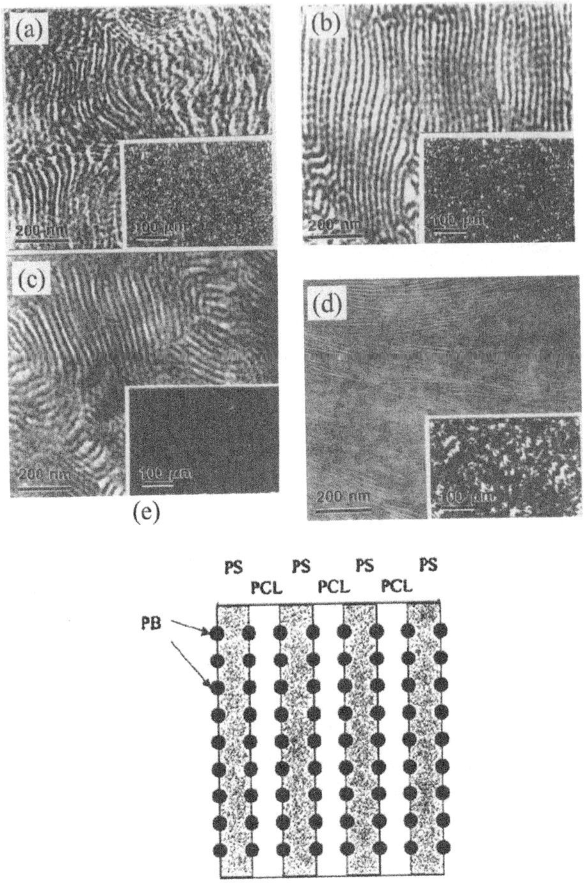

Fig. 15 Microphase morphology of $S_{35}B_{15}C_{50}$ triblock copolymer bulk films (**a**, **b**), thin film (**c**, **d**) by TEM with *insets* of polarized optical microscopy (*POM*): (**a**, **c**) as cast; (**b**, **d**) after thermal treatment, TEM reveals a lamellar cylindrical structure, with evidence of enhanced long-range order upon annealing above the T_g of PS. For all TEM images presented, the *white areas* are PCL-rich lamellae, the *gray areas* are PS-rich lamellae and the *black areas* represent PB-rich cylinders after staining with OsO_4. The POM images show a speckle pattern indicating local crystallization of PCL in the microphase-separated structure. (**e**) Schematic representation of lamellar–cylindrical morphology for the $S_{35}B_{15}C_{50}$ triblock copolymer in cross section, where the *white areas* are PCL-rich lamella, the *gray areas* are PS-rich lamella and the *black areas* represent PB-rich cylinders. (Reprinted with permission from [125]. Copyright 2001 American Chemical Society)

crystallization behavior between bulk and thin films may be that the T_g of PS is shifted to lower temperature owing to an incomplete phase separation in the thin films, perhaps because of a faster cooling rate and concomitant partial miscibility between the components. Nevertheless, what is interesting here is that it does not seem possible to have spherulitic superstructures along with a long-range ordered microphase-separated morphology. Similar results have been reported by Balsamo et al. [122]. In fact, agreement between calculated Avrami parameters and morphological observations has been found.

Ueda et al. [26] recently investigated a flow-oriented PE-*b*-aPP diblock copolymer with $M_w = 113\,000$ ($M_n/M_w = 1.1$) and a PE volume fraction of 0.48. This diblock copolymer is in the strong segregation regime (i.e., estimated $\chi N = 10.5$ and $T_{ODT} = 290\,°C$) and has a lamellar morphology in the melt. They found a breakout phenomenon with the formation of spherulites in an intermediate crystallization temperature range: $95 \leq T_c \leq 101\,°C$. At crystallization temperatures above 101 °C or below 95 °C spherulites were not formed and the crystallization was confined within the lamellar MD. Ueda et al. report that lamellar MD and spherulites do not co-exist when the material crystallizes from the melt which is separated in lamellar MDs. In other words, in this particular case, breakout or confined crystallization within lamellar MDs depends on the crystallization conditions.

Laredo et al. [131–133] performed thermally stimulated depolarization current (TSDC) experiments on a series of PS-*b*-PB-*b*-PCL triblock copolymers with varying amounts of PCL. The relaxation time distribution was extracted by a numerical decomposition of the TSDC spectra and it was shown that this distribution was independent of the PCL content on going from the homopolymer to the triblock copolymer with 16 to 77 wt % of PCL, in samples that were not annealed and were employed as synthesized. Better segregation of the mesophase structure was reached when the samples were submitted to annealing at 140 °C and important variations in the TSDC and WAXS spectra were obtained as a result of the thermal treatment. For the $S_{09}B_{14}C_{77}$ triblock copolymer the results obtained were explained by postulating the existence of a rigid amorphous phase in the PCL block. Such a rigid amorphous phase could be located between the core–shell cylinders formed by the other blocks with PS as the core and PB as the shell and may be constrained by undulated lamellae of crystalline PCL. The results for the material with the less abundant PCL block are explained as a result of the confinement in nanotubes of PCL surrounded by PB embedded in a vitreous PS matrix. Broadband dielectric experiments on these same materials confirmed the results obtained by TSDC spectroscopy.

6
ABC Triblock Copolymers with Two Crystallizable Blocks

Only a few publications deal with ABC triblock copolymers where two of the blocks are able to crystallize. The systems that have been investigated include PS-*b*-PE-*b*-PCL [94, 98], PE-*b*-PS-*b*-PCL [94], PS-*b*-PEO-*b*-PCL [30, 134–136] and PE-*b*-poly(ethylene-propylene)-*b*-PEO [101, 119] (see also Table 1).

6.1 Influence of Composition and Crystallizable Block Position within ABC Triblock Copolymers

The influence of composition has been studied and its influence on the crystallization behavior can be very important. Arnal et al. [30, 134] demonstrated that when PS-*b*-PEO-*b*-PCL triblock copolymers are cooled from the melt, crystallization at very different supercoolings is observed depending on the amount of PCL in the copolymer. The effect of increasing PCL content in the crystallization and melting behavior of PS-*b*-PEO-*b*-PCL triblock copolymer is shown in Fig. 16a. An AB diblock precursor, $S_{81}EO_{19}{}^{18.5}$, has been included for comparison purposes. It can be seen that its crystallization takes place at very low temperatures (i.e., – 40 °C). The PEO block in this case crystallizes from homogeneous nuclei as a result of its confinement in a vast number of small and isolated MDs in a vitreous PS matrix. The fact that homogeneous nucleation of PEO was obtained can be inferred from the vicinity of the crystallization temperature to the T_g of the PEO block, which is found at – 58 °C. The T_g of the PS block can be clearly seen in the subsequent heating scan and is therefore consistent with a system that is phase-segregated in the melt (Fig. 16b).

For a $S_{63}EO_{16}C_{21}{}^{24}$ triblock copolymer, a vitreous PS matrix contains minor PCL and PEO contents. It can be deduced from the crystallization behavior from the melt that both PCL and PEO crystallize at maximum su-

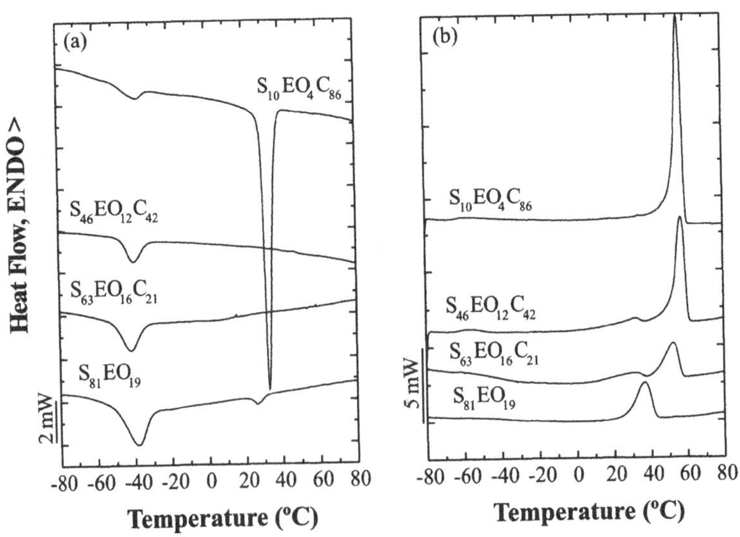

Fig. 16 DSC **a** cooling and **b** heating scans (10 °C min^{-1}) of the PS-*b*-PEO diblock precursor and PS-*b*-PEO-*b*-PCL triblock copolymer. (Reprinted with permission from [30]. Copyright 2001 American Chemical Society)

percooling and in a coincident fashion at – 42 °C. This means that both minor components, the PEO and the PCL, are homogeneously nucleated. No crystallization exotherms were observed at the usual crystallization temperatures of PCL and PEO homopolymers; this is a consequence of the total isolation of the well-segregated MDs [29, 30, 99, 134]. On the other hand, Arnal et al. have shown that when the composition is changed, triblock copolymers like $S_{46}EO_{12}C_{42}{}^{33}$ and $S_{10}EO_{4}C_{86}{}^{150}$ can exhibit a fractionated crystallization behavior with at least two distinct crystallization exotherms. In both cases, a low-temperature exotherm is present which could be interpreted as the crystallization of isolated PEO MDs. Then, there are high-temperature exotherms, at about 27 and 33 °C, respectively, which should correspond to PCL crystallization. However, in view of the small size of the high-temperature exotherm corresponding to the PCL component in the $S_{46}EO_{12}C_{42}{}^{33}$ triblock copolymer, it is possible that part of the PCL component crystallizes in a fractionated fashion at coincident temperatures with the PEO component at around – 40 °C. This result indicates that not all the PCL MDs are percolated and that a certain number of PCL MDs are isolated and may not contain the highly active heterogeneities that induce crystallization at lower supercoolings. For the measurements presented in Fig. 16, the triblock copolymers were not submitted to long annealing processes to perfect morphology and this may contribute to their fractionated crystallization behavior [30].

The DSC heating scans presented in Fig. 16b show two melting endotherms for all triblock copolymers. From a comparison with the $S_{81}EO_{19}$ precursor, it can be deduced that the endotherm located at lower temperatures is due to the melting of PEO. Nevertheless, the melting peak temperatures of the low-temperature endotherms of the triblock copolymers are shifted to lower temperatures (30, 29 and 20 °C for the triblock copolymers versus 35 °C for the diblock precursor). This can be attributed to increased topological restrictions in the PEO block as a consequence of the chemical link with another crystallizable block (PCL) and the resulting morphology. The comparison of the crystallinity degrees between the diblock precursor and triblock copolymers corroborates this explanation, since the PEO blocks in the triblock cases have lower crystallinity values ($X_c = 39\%$ for $S_{81}EO_{19}$ and $X_c = 16\%$ for $S_{46}EO_{12}C_{42}$).

Several of the ABC triblock copolymers with two crystallizable blocks that have been studied include PE as one of the crystallizable components. The PE block can be found either at the end (PE-*b*-PS-*b*-PCL [94], PE-*b*-PEP-*b*-PEO [101, 119]) or at the center (PS-*b*-PE-*b*-PCL [98]). When the PE block is located at the center of the copolymer, as is the case in PS-*b*-PE-*b*-PCL triblock copolymers [94], there are higher constraints on the PE block owing to the absence of free ends. If the PE block is a minor component, confined crystallization with possible homogeneous nucleation is usually encountered. It may be possible that when the PE block does not have free ends, it may be more difficult for its crystallization to induce a breakout with a concomitant MD percolation than when it is located at the ends of the triblock copolymer. As a consequence, the crystallization temperature and the melting point of the PE block in PS-*b*-PE-*b*-PCL decrease as the confinement degree increases. Thus, $S_{57}E_{27}C_{16}$ and $S_{36}E_{35}C_{29}$ (whose precursors had PB blocks with 12% by weight of 1.2 units) melt at 85.1 and 89.5 °C, respectively (about 5–10 °C less than in PE-*b*-PS-*b*-PCL). Nevertheless, it should be mentioned that even in cases where the PE block is not morphologically confined, a slight reduction of the melting and crystallization temperatures with respect to hydrogenated PB (PE) can be observed. Therefore, it can be concluded that the combined effects of confined crystallization and topological constraints due to the lack of free ends are responsible for the reduction of the crystallization temperature of the PE block in this case. Moreover, it has additionally been observed that the crystallinity degree of the PE block is significantly depressed when compared with that of neat hydrogenated PB, and these effects can also influence the crystallization kinetics.

PE-*b*-PEP-*b*-PEO triblock copolymers [101, 119] are very interesting because χ_{PE-PEO} and $\chi_{PEP-PEO}$ are higher than χ_{PE-PEP}. As a consequence of the strong incompatibility of the polar PEO segments with respect to the other block components, differences in the degree of confinement between the crystallizable blocks are observed that affect the process of crystallization. In $E_{11}EP_{71}EO_{18}{}^{123}$ the PEO is confined into isolated MDs and crystallization takes places at temperatures near – 25 °C in a fractionated way. Although the crystallization temperature is very low compared with that of PEO homopolymer (T_c = 20–40 °C), it is not low enough to originate from homogeneous nucleation (Sect. 5.1), and possible surface nucleation effects have been postulated by Schmalz et al. [101] to explain the origin of such a low-temperature exotherm. Self-nucleation experiments confirm that PEO MDs are isolated and domain II disappears for the PEO that crystallizes at – 25 °C (Sect. 6.3). This behavior is observed in PE-*b*-PEP-*b*-PEO copolymers with PEO blocks with compositions between 18 and 41%. For the PE blocks, within PE-*b*-PEP-*b*-PEO triblock copolymers, the situation is different. Owing to the low segmental interaction parameter between PEP and PE, the crystallization of PE chains occurs from a homogeneous mixture of PE and PEP chains without confinement, for example, in $E_{18}EP_{57}EO_{25}{}^{133}$, the PE block crystallizes at 64 °C. Similar behavior was observed for PE blocks in a range between 11 and 24%. The crystallization of the PE block was nearly independent of the triblock composition and always occurred at temperatures close to those exhibited by neat hydrogenated PB. Crystallization is induced by heterogeneous nucleation and spreads over the entire PE domain. The morphological study of $E_{19}EP_{40}EO_{41}$ shows a continuous semicrystalline PE domain consisting of interconnected PE crystallites in an hexagonal array with crystalline PEO cylinders that are embedded in a matrix of the PEP block. PEO cylinders exhibit a rectangular shape which might be attributed to the fact that the PEO cylinders are semicrystalline (Fig. 13b, c).

Important differences in crystallization behavior were observed between the two crystallizable blocks within PE-*b*-PEP-*b*-PEO triblock copolymers. The crystallization of PE occurs without confinement from a homogeneous mixture of PE and PEP segments, but for PEO blocks, the crystallization takes place exclusively at high supercoolings in small isolated MDs.

6.2 Self-Nucleation Behavior

The technique of self-nucleation [75] can be very useful to study the nucleation and crystallization of block copolymer components, as already mentioned in previous sections. In block copolymers, factors like the volumetric fraction and the degree of segregation affect the type of confinement and therefore modify the self-nucleation behavior. In the case of semicrystalline block copolymers, several works have reported the self-nucleation of either one or both crystallizable components in PS-*b*-PCL, PS-*b*-PB-*b*-PCL, PS-*b*-PE-*b*-PCL, PB-*b*-PIB-*b*-PEO, PE-*b*-PEP-*b*-PEO, PS-*b*-PEO, PS-*b*-PEO-*b*-PCL, PB-*b*-PEO, PB/PB-*b*-PEO and PPDX-*b*-PCL [29, 92, 98, 99, 101–103, 134] and three different kinds of behavior have been observed. Specific examples of these three cases are given in the following and in Table 5:

1. The "classical" behavior is observed when the block under study is a continuous phase or a percolated MD and crystallization is induced by heterogeneous nucleation, because the probability of a highly active heterogeneity being located in the crystallizable domain is sufficiently high. It is also observed in block copolymers whose crystallization drives structure formation, i.e., when they crystallize from a homogeneous melt or from a weakly segregated melt. The PE block within PE-*b*-PEP-*b*-PEO triblock copolymers is an example of a block that crystallizes from a melt mixed PEP/PE phase; self-nucleation experiments in $E_{18}EP_{57}EO_{25}{}^{133}$ showed that domain II was reached at a T_s of 99 °C. Upon a further decrease of T_s, annealing takes place at 95 °C (Fig. 17a). In PE-*b*-PEP-*b*-PEO triblock copolymers domain II is always present for the PE blocks, regardless of the PE content between 10 and 25% [101, 119]. In weakly segregated systems, like PPDX-*b*-PCL, heterogeneous nucleation events that start within the MD spread by secondary nucleation once the breakout phenomenon that allows crystallization to drive structure formation sets in [102, 103].
2. Several block copolymer systems have shown only domains I and III upon self-nucleation. This behavior is observed in confined crystallizable blocks as PEO in purified $E_{24}EP_{57}EO_{19}{}^{69}$ [29]. Crystallization takes place for the PEO block at − 27 °C after some weak nucleating effect of the interphase. Domain II is absent and self-nucleation clearly starts at $T_s = 56$ °C when annealed crystals are already present, i.e., in domain III (Fig. 17b). The absence of domain II is a direct consequence of the extremely high

Table 5 Self-nucleation behavior for diblock and triblock copolymers

System	Self-Nucleation Domains 1st Crystallizable Block	Self-Nucleation Domains 2nd Crystallizable Block	Ref.
$B_{24}I_{56}EO_{20}$	I/III_{SA}		101
$B_{11}I_{70}EO_{19}$	I/III_{SA}		101
$B_{17}I_{57}EO_{26}$	I/III_{SA} and $I/II/III_{SA}$		101
$B_{19}I_{39}EO_{42}$	I/III_{SA} and $I/II/III_{SA}$		101
PB_{16}/B_{50}-*b*-EO_{50}	I/II/III		92
$E_{24}EP_{57}EO_{19}$	I/II/III	I/III_{SA} and I/II/III	101
$E_{24}EP_{57}EO_{19}$ (purified)	I/II/III	I/III_{SA}	101
$E_{11}EP_{71}EO_{18}$	I/II/III	I/III_{SA} and I/II/III	101
$E_{11}EP_{71}EO_{18}$ (purified)	I/II/III	I/III_{SA}	101
$E_{18}EP_{57}EO_{25}$	I/II/III	I/III_{SA}	101
$E_{19}EP_{40}EO_{41}$	I/II/III	$I/III_{A}/III_{SA}$ and I/II/III	101
$S_{81}EO_{19}$	$I/III_{A}/III_{SA}$		99
$S_{39}EO_{61}$	I/II/III		99
$S_{63}EO_{16}C_{21}$	$I/III_{A}/III_{SA}$	$I/III_{A}/III_{SA}$	134
$S_{15}EO_{37}C_{48}$	I/II/III	I/II/III	134
$S_{27}C_{73}$	I/III_{SA} and $I/II/III_{SA}$		29
$S_{62}B_{27}C_{11}$	$I/III_{A}/III_{SA}$		29
$S_{62}B_{36}C_{38}$	I/III_{SA} and I/II/III		29
$S_{27}E_{15}C_{58}$	$I/III_{A}/III_{SA}$		98
$S_{35}E_{15}C_{50}$	$I/III_{A}/III_{SA}$		98
$S_{57}E_{27}C_{16}$	I/III_{SA}		98
$S_{27}E_{23}C_{36}$	I/II/III		98
$D_{23}{}^{8}C_{77}{}^{27}$	I/II/III	I/II/III	102
$D_{28}{}^{10}C_{72}{}^{24}$	I/II/III	I/II/III	102
$D_{40}{}^{5}C_{60}{}^{7}$	I/II/III	I/II/III	102
$D_{55}{}^{7}C_{45}{}^{6}$	I/II/III	I/II/III	102
$D_{77}{}^{32}C_{23}{}^{10}$	I/II/III	I/II/III	102

density of MD structures that need to be self-seeded (of the order of $10^{15}-10^{16}$ cm^{-3}). Therefore, to increase the density of self-nuclei, the self-nucleation temperature has to be decreased to values so low that some unmelted crystals can be annealed. It must be stressed that only in cases where the nucleation is homogeneous or superficial is this type of behav-

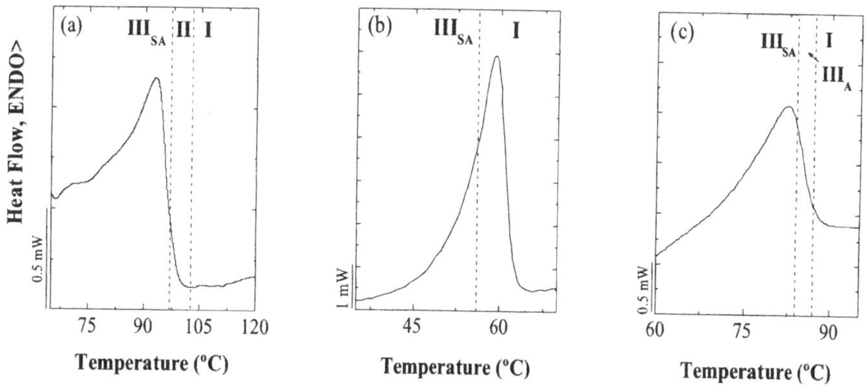

Fig. 17 **a** Classical self-nucleation behavior for polyethylene (*PE*) within $E_{18}EP_{57}EO_{25}{}^{133}$ triblock copolymer [101]. **b** Self-nucleation behavior for PEO within purified $E_{24}EP_{57}EO_{19}{}^{64}$ triblock copolymer [29]. **c** Self-nucleation behavior for PE within $S_{35}E_{15}C_{50}{}^{219}$ triblock copolymer [29,98]. (**a**, **c** from [98,101] with permission. **b** Reprinted with permission from [29]. Copyright 2002 American Chemical Society)

ior observed. When low crystallization exotherms are produced by any type of heterogeneity, domain II is always observed [101].

3. In those cases where the injection of self-nuclei in every MD is most difficult in view of the very large number of MDs, domain III is split into two domains. Evaluation of the self-nucleation of the PE block within $S_{35}E_{15}C_{50}{}^{219}$ shows that not only domain II is absent, but upon decreasing T_s, the PE block is annealed before any detectable self-nucleation occurs (Fig. 17c). Therefore two subdomains were defined [98]: domain III_A, where annealing without previous self-nucleation occurs and domain III_{SA}, where self-nucleation and annealing are simultaneously observed for $T_s < 88$ °C. Domain III_{SA} would be the exact equivalent to the standard domain III established by Fillon et al. [75].

Several systems exhibited mixed behavior (Table 5) because crystallization occurs in a fractionated way (e.g., $B_{19}I_{39}EO_{42}$). The fraction of crystals formed at higher temperatures exhibits the three classical nucleation domains, and those crystals that crystallize at the largest supercoolings can only be self-nucleated at temperatures where domain III has already started and therefore domain II disappears for this population. In semicrystalline diblock or triblock copolymers, when fractionated crystallization is observed with more than one exothermal signal, the high-temperature exotherm may be attributed to heterogeneous crystallization of percolated domains and the low-temperature one to crystallization in isolated domains, depending on the specific crystallization temperature. An independent proof of the isolated nature of the MD is required [96] besides dynamic DSC evidence (either a morphological confirmation or a crystallization kinetics study that yields

Avrami indexes of the order of 1 or lower) unless the crystallization occurs at temperatures very close to the T_g of the crystallizing phase.

Table 5 presents results of the self-nucleation behavior of PE, PEO, PPDX and PCL blocks in 27 block copolymers. A general trend can be observed in AB and ABC block copolymers as outlined before. Heterogeneity-free MDs will nucleate either superficially or homogeneously as expected. However, domain II will not be observed as a result of the great number of self-seeds that these crystallizable blocks demand in order to be self-nucleated since they are confined into MDs like cylinders or spheres where the density of such regions can be as high as $10^{15} - 10^{16}$ cm^{-3}. This means that the number of MDs exceeds the number of active heterogeneities by several orders of magnitude. It is well known that as T_s decreases, the nucleation density sharply increases as self-nuclei are produced in the material. In these cases of crystallizable blocks within confined MDs, T_s needs to be lowered to such a degree that partial melting occurs extensively, and annealing is produced simultaneously or even before self-nucleation can be realized in every MD.

Chen et al. [92] also performed self-nucleation experiments by DSC in PB-*b*-PEO diblock copolymers and PB/PB-*b*-PEO blends. The cooling scans presented in their work showed that a classical self-nucleation behavior was obtained for PEO homopolymer and for the PB/PB-*b*-PEO blend where the weight fraction of PEO was 0.64 and the morphology was lamellar in the melt. For PB/PB-*b*-PEO blends with cylinder or sphere morphology, the crystallization temperature remained nearly constant for several self-seeding temperatures evaluated. This observation indicates that domain II or the self-nucleation domain was not observable for these systems, as expected in view of the general trend outlined earlier.

The study of coincident crystallization with self-nucleation techniques was already mentioned in Sect. 4 in the case of the weakly segregated PPDX-*b*-PCL diblock copolymer system. Self-nucleation experiments can also be used to study triblock copolymers that exhibit coincident crystallization of two of the constituent blocks upon cooling from the melt in a differential scanning calorimeter, as happens in $S_{63}EO_{16}C_{21}{}^{24}$ and $S_{15}EO_{37}C_{48}{}^{64}$ (Fig. 16). The self-nucleation technique can also be employed to gain a separation of the crystallization and/or melting signal of each block within ABC triblock copolymers with two crystallizable blocks when coincident transitions are observed in standard DSC runs [99, 102, 103, 134].

6.3 Influence of Copolymer Architecture: Star Versus Linear Triblock Copolymers

Molecular architecture modifies the phase behavior of block copolymers. In block copolymers, macrophase separation is prevented by the connectivity of the polymer chains. The transition from a homogeneous melt to a heterogeneous melt with ordered MDs is called the ODT. It occurs at a critical value of χN depending on composition and architecture. Theoretical and experimental studies reveal a general trend as follows [3]:

$$(\chi N)_{c,\,\text{blend}} < (\chi N)_{\text{ODT, diblock}} < (\chi N)_{\text{ODT, graft}} < (\chi N)_{\text{ODT, triblock}} < (\chi N)_{\text{ODT, star}}.$$

Star and linear PS-*b*-PEO-*b*-PCL triblock copolymers were studied by Floudas et al. [135, 136] and Arnal et al. [30, 134]. Crystallization in the star block copolymers started from a homogeneous melt and therefore crystallization drove structure formation as indicated by SAXS and WAXS results. The linear ABC triblock copolymer had different behavior; DSC evaluation detected a single glass-transition temperature at about − 60 °C that represents the overlap of the PCL and PEO T_gs and a high-temperature glass transition at about 90 °C, corresponding to the PS block, indicating that the block copolymers are microphase-separated. A SAXS experiment of the melt for some compositions confirmed melt segregation. The observed differences in the degree of segregation between star and linear triblock copolymers are in agreement with theoretical studies [3]. Also the molecular weights of the PS-*b*-PEO-*b*-PCL star block copolymers are lower than those of the linear triblock copolymers; such a relationship between their respective molecular weights enhances the differences in the behavior between the star and the linear PS-*b*-PEO-*b*-PCL triblock copolymers.

Fractionated crystallization at high supercoolings was observed when either one or both blocks constituted the minor components in linear triblock copolymers (as shown in Fig. 16a). Four star triblock copolymers were evaluated by Floudas et al. [135, 136]: $S_{18}EO_{75}C_7{}^{27}$, $S_{13}EO_{58}C_{29}{}^{35}$, $S_7EO_{29}C_{64}{}^{70}$ and $S_4EO_{18}C_{78}{}^{112}$. The star triblock copolymers were crystallized isothermally at 39–44 °C, and only $S_{13}EO_{58}C_{29}{}^{35}$ and $S_7EO_{29}C_{64}{}^{70}$ exhibited crystallization for both PEO and PCL blocks. In $S_{18}EO_{75}C_7{}^{27}$ only the PEO block crystallizes and in $S_4EO_{18}C_{78}{}^{112}$ only PCL crystallizes. Floudas et al. [135, 136] proposed that the length ratio between crystallizable blocks affects crystallization and only when this ratio is less than 3 both blocks are able to crystallize. In very asymmetric stars (i.e., $S_{18}EO_{75}C_7{}^{27}$ and $S_4EO_{18}C_{78}{}^{112}$), Floudas et al. proposed that the longer block was found to completely suppress the crystallization of the shorter one. In our opinion, it would be convenient to explore the crystallization behavior for these PS-*b*-PEO-*b*-PCL star triblock copolymers at higher supercoolings, since maybe at much lower temperatures the shorter block could crystallize.

The study of both star and linear PS-*b*-PEO-*b*-PCL triblock copolymers demonstrates the complexity of the crystallization behavior of ABC triblock copolymers and also the multiple possibilities of modifying the crystallization behavior of the block components by changing composition and/or molecular architecture.

6.4 Influence of Thermal Treatments on Nonisothermal and Isothermal Crystallization

When PS-*b*-PE-*b*-PCL triblock copolymers were investigated [98], it was observed that if the copolymer contained small amounts of PCL, this component exhibited fractionated crystallization upon cooling from the melt even though the PE block had already crystallized at higher temperatures. In addition, a general decrease of the crystallization temperature of the PCL component within PS-*b*-PE-*b*-PCL triblock copolymers compared with that within PS-*b*-PB-*b*-PCL was observed (e.g., 20 °C in SBC vs. 14 °C in SEC). This was the first evidence to suggest that there was no nucleation effect of the semicrystalline PE block on the crystallization of the PCL block. On the basis of the self-nucleation domains found for the PCL homopolymer, self-nucleation experiments were carried out in the SEC triblock copolymers and compared with SBC to asses the effect of the previous crystallization of the PE block. The self-seeding temperature was varied from 140 to 50 °C. Figure 18 shows a comparison between $S_{27}B_{15}C_{58}{}^{219}$ and $S_{27}E_{15}C_{58}{}^{219}$. It can be observed that the PCL crystallization temperatures were depressed in the SEC when the T_s employed were in the lowest melting temperature tail of the PE block. The phenomenon was called an "antinucleation effect," since a lowering of the crystallization temperature is obtained. Because this effect is absent in the PS-*b*-PB-*b*-PCL analogue copolymer, it was ascertained that this effect is a consequence of the annealing of the thinnest or the more defective PE crystals. Such crystals may be the ones closer to the interphase between PCL and PE.

Recently, the isothermal crystallization kinetics of the PE block within SEC triblock copolymers was investigated [94]. When the PE content in the

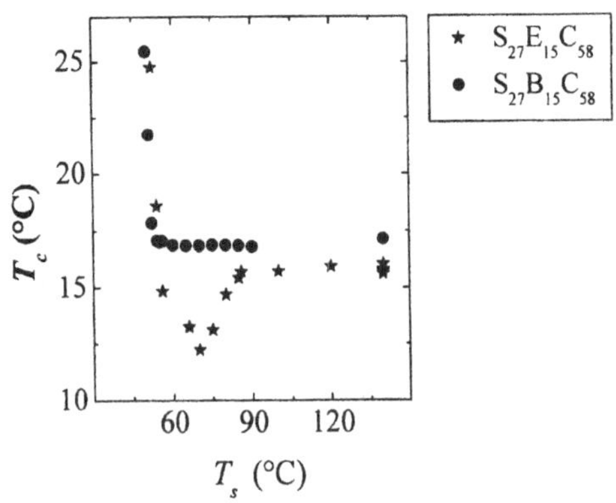

Fig. 18 Peak crystallization temperature as a function of self-nucleation temperature for hydrogenated and nonhydrogenated $S_{27}B_{15}C_{58}$ triblock copolymers. (Reprinted with permission from [97]. Copyright 1998 American Chemical Society)

copolymers was equal to or lower than 40%, it was very difficult to detect the exothermal heat of crystallization during isothermal DSC experiments. The authors employed an indirect technique to determine the crystallization kinetics by DSC. Melting scans after isothermal crystallization performed at different times were employed to determine the crystallization kinetics one step at a time ("isothermal step crystallization"). Double-melting endotherms were observed after isothermal crystallization and they were interpreted as a result of the melting of two lamellar populations. As the degree of PE confinement increased, the Avrami index decreased to values that are even lower than 1. Figure 19 shows the calculated Avrami indexes for the PE block within a series of SEC triblock copolymers. It is evident that the Avrami index decreases as the PE content in the SEC triblock copolymers decreases, except for the ESC, where the PE block is located at the end. This indicates that there is a clear correlation between a decrease of the Avrami index and an increase in the confinement degree of the PE block within the copolymers. In the cases where PE confinement is high, a probable homogeneous nucleation (or surface nucleation) was detected (through self-nucleation experiments), and the Avrami indexes were very low, 0.5 or lower (Fig. 19). Such extremely low values can be explained by the nature of the homogeneous nucleation process that in this case is between sporadic and instantaneous (Sect. 3, Eq. 8 and its explanation). Finally, it should mentioned that these materials show an increase in the supercooling needed for crystallization as the degree of confinement for the PE block increases and that simultaneously there is a concomitant increase in the overall transformation rate.

Floudas et al. [135] also studied the isothermal crystallization of PEO and PCL blocks within PS-*b*-PEO-*b*-PCL star triblock copolymers. In these systems the crystallization occurs from a homogeneous melt; Avrami indexes higher than 1 are always observed since the crystallization drives structure formation and does not occur under confined conditions. A reduction in the equilibrium melting temperature in the star block copolymers was also observed.

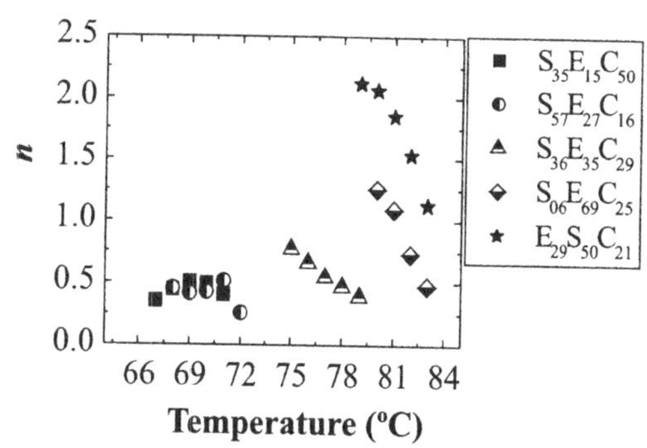

Fig. 19 Avrami index as a function of crystallization temperature for PS-*b*-PE-*b*-PCL and PE-*b*-PS-*b*-PCL triblock copolymers. (Reprinted from [94] with permission from Elsevier)

Acknowledgements We would like to express our gratitude to the late Prof. Reimund Stadler for introducing us into the world of block copolymers. Special thanks are due to Volker Abetz for many enlightening discussions in the framework of our collaboration on the subject of block copolymer crystallization. A.J.M. would like to acknowledge interesting discussions on the subject of droplet crystallization with G. Reiter and K. Dalnoki-Veress and early disclosure of unpublished work. We would also like to thank two of our present M.Sc. students, Julio Albuerne and Arnaldo Lorenzo, for their invaluable help in the compilation and preparation of references, tables and figures.

References

1. Hamley IW (1998) The physics of block copolymers. Oxford University Press, Oxford
2. Alexandridis P, Holzwarth JF (2000) Curr Opin Colloid Interface Sci 5:312
3. Hadjichristidis N, Pispas S, Floudas GA (2003) Block copolymers. Wiley, New York
4. Riess G, Hurtrez G, Bahadur P (1985) In: Mark HF, Kroschwitz JI (eds) Encyclopedia of polymer science and engineering, vol 2. Wiley, New York, p 324
5. Hamley IW (2004) (ed) Developments in block copolymer science and technology. Wiley, New York
6. Rangarajan P, Register RA, Fetters LJ (1993) Macromolecules 26:4640
7. Rangarajan P, Register RA, Adamson DH, Fetters LJ, Bras W, Naylor S, Ryan AJ (1995) Macromolecules 28:1422
8. Ryan AJ, Hamley IW, Bras W, Bates FS (1995) Macromolecules 28:3860
9. Richardson PH, Richards RW, Blundell DJ, MacDonald WA, Mills P (1995) Polymer 36:3059
10. Quiram DJ, Register RA, Marchand GR, Ryan AJ (1997) Macromolecules 30:8338
11. Douzinas KC, Cohen RE, Halasa AF (1991) Macromolecules 24:4457
12. Hamley IW, Patrick J, Fairclough A, Terrill NJ, Ryan AJ, Lipic PM, Bates FS, Towns-Andrews E (1996) Macromolecules 29:8835
13. Rangarajan P, Register RA, Fetters LJ, Bras W, Naylor S, Ryan AJ (1995) Macromolecules 28:4932
14. Nojima S, Kato K, Yamamoto S, Ashida T (1992) Macromolecules 25:2237
15. Quiram DJ, Register RA, Marchand GR (1997) Macromolecules 30:4551
16. Rohadi A, Endo R, Tanimoto S, Sasaki S, Nojima S (2000) Polym J 32:602
17. Rohadi A, Tanimoto S, Sasaki S, Nojima S (2000) Polym J 32:859
18. Ryan AJ, Fairclough JPA, Hamley IW, Mai SM, Booth C (1997) Macromolecules 30:1723
19. Zhang F, Chen Y, Huang H, Hu Z, He T (2003) Langmuir 19:5563
20. Hillmyer MA, Bates FS (1997) Macromol Symp 117:121
21. Loo YL, Register RA, Ryan AJ (2002) Macromolecules 35:2365
22. Nojima S, Toei M, Hara S, Tanimoto S, Sasaki S (2002) Polymer 43:4087
23. Reiter G, Castelein G, Sommer JU, Röttele A, Thurn-Albrecht T (2001) Phys Rev Lett 87:226101
24. Robitaille C, Prud'homme J (1983) Macromolecules 16:665
25. Quiram DJ, Register RA, Marchand GR, Adamson DH (1998) Macromolecules, 31:4891
26. Ueda M, Sakurai K, Okamoto S, Lohse D, MacKnight WJ, Shinkai S, Sakurai S, Nomura S (2003) Polymer 44:6995
27. Reiter G (2003) J Polym Sci Part B Polym Phys 41:1869

28. Reiter G, Castelein G, Sommer JU (2003) In: Sommer JU, Reiter G (eds) Polymer crystallization, observations, concepts and interpretations. Springer, Berlin Heidelberg New York, p 131
29. Müller AJ, Balsamo V, Arnal ML, Jakob T, Schmalz H, Abetz V (2002) Macromolecules 35:3048
30. Arnal ML, Balsamo V, López-Carrasquero F, Contreras J, Carrillo M, Schmalz H, Abetz V, Laredo E, Müller AJ (2001) Macromolecules 34:7973
31. Floudas G, Tsitsilianis C (1997) Macromolecules 30:4381
32. Gervais M, Gallot B (1981) Polymer 22:1129
33. Lotz B, Kovacs AJ (1969) Polym Prepr Am Chem Soc Div Polym Chem 10:820
34. O'Malley JJ (1977) J Polym Sci Polym Symp 60:151
35. Zhu L, Chen Y, Zhang A, Calhoun BH, Chun M, Quirk RP, Cheng SZD, Hsiao BS, Yeh F, Hashimoto T (1999) Phys Rev B 60:10022
36. Zhu L, Cheng SZD, Calhoun BH, Ge Q, Quirk RP, Thomas EL, Hsiao BS, Yeh F, Lotz B (2001) Polymer 42:5829
37. Zhu L, Mimnaugh BR, Ge Q, Quirk RP, Cheng SZD, Thomas EL, Lotz B, Hsiao BS, Yeh F, Liu L (2001) Polymer 42:9121
38. Zhu L, Cheng SZD, Huang P, Ge Q, Quirk RP, Thomas EL, Lotz B, Hsiao BS, Yeh F, Liu L(2002) Adv Mater 14:31
39. Zhu L, Huang P, Chen WY, Ge Q, Quirk RP, Cheng SZD, Thomas EL, Lotz B, Hsiao BS, Yeh F, Liu L (2002) Macromolecules 35:3553
40. Xu JT, Yuan JJ, Cheng SY (2003) Eur Polym J 39:2091
41. Weimann PA, Hajduk DA, Chu C, Chaffin KA, Brodil JC, Bates FS (1999) J Polym Sci Part B Polym Phys 37:2053
42. Loo YL, Register RA, Ryan AJ, Dee GT (2001) Macromolecules 34:8968
43. Hamley IW (1999) Adv Polym Sci 148:114
44. Loo YL, Register RA (2004) In: Hamley IW (ed) Developments in block copolymer science and technology. Wiley, New York, p 213
45. Mogi Y, Nomura M, Kotsuji H, Ohnishi K, Matsushita Y, Noda I (1994) Macromolecules 27:6755
46. Stadler R, Auschra C, Beckmann J, Krappe U, Voigt-Martin IG, Leibler L (1995) Macromolecules 28:3080
47. Breiner U, Krappe U, Thomas EL, Stadler R (1998) Macromolecules 31:135
48. Bates FS, Fredrickson GH (1999) Phys Today 52:32
49. Hückstadt H, Goldacker T, Göpfer A, Abetz V (2000) Macromolecules 33:3757
50. Abetz V, Goldacker T (2000) Macromol Rapid Commun 21:16
51. Nakazawa H, Otha T (1993) Macromolecules 26:5503
52. Kane L, Spontak RJ (1994) Macromolecules 27:663
53. Zheng W, Wang ZG (1995) Macromolecules 28:7215
54. Breiner U, Krappe U, Abetz V, Stadler R (1997) Macromol Chem Phys 198:1051
55. Matsen MW (1998) J Chem Phys 108:785
56. Abetz V, Stadler R, Leibler L (1996) Polym Bull 37:135
57. Vonnegut B (1948) J Colloid Sci 3:563
58. Turnbull D, Cech RE (1950) J Appl Phys 21:804
59. Pound GM, La Mer VK (1952) J Am Chem Soc 74:2323
60. Turnbull D (1952) J Chem Phys 20:411
61. Turnbull D, Cormia RL (1961) J Chem Phys 34:820
62. Cormia RL, Price FP, Turnbull D (1962) J Chem Phys 37:1333
63. Burns JR, Turnbull D (1966) J Appl Phys 37:4021
64. Koutsky JA, Walton AG, Baer E (1967) J Appl Phys 38:1832

65. Gornick F, Ross GS, Frolen LJ (1967) J Polym Sci C18:79
66. Barham PJ, Jarvis DA, Keller A (1982) J Polym Sci Polym Phys Ed 20:1733
67. Frensch H, Harnischfeger P, Jungnickel BJ (1989) In: Utracky LA, Weiss RA (eds) Multiphase polymers: blends and ionomers. ACS symposium series 395. American Chemical Society, Washington. DC, p 101, and references therein
68. Arnal ML, Matos ME, Morales RA, Santana OO, Müller AJ (1998) Macromol Chem Phys 199:2275
69. Santana OO, Müller AJ (1994) Polym Bull 32:471
70. Morales RA, Arnal ML, Müller AJ (1995) Polym Bull 35:379
71. Manaure AC, Morales RA, Sánchez JJ, Müller AJ (1997) J Appl Polym Sci 66:2481
72. Arnal ML, Müller AJ (1999) Macromol Chem Phys 200:2559
73. Manaure A, Müller AJ (2000) Macromol Chem Phys 201:958
74. Arnal ML, Müller AJ, Maiti P, Hikosaka M (2000) Macromol Chem Phys 201:2493
75. Fillon B, Wittman JC, Lotz B, Thierry A (1993) J Polym Sci Part B Polym Phys 31:1383
76. Price FP (1969) In: Zettlemoyer AC (ed) Nucleation. Dekker, New York, p 405
77. Talsma H, van Steenbergen MJ, Crommelin DJA (1992) Cryobiology 29:80
78. Janssen AH, Talsma H, van Steenbergen MJ, de Jong KP (2004) Langmuir 20:41
79. Montenegro R, Landfester K (2003) Langmuir 19:5996
80. Montenegro R, Antonietti M, Mastai Y, Landfester K (2003) J Phys Chem B 107:5088
81. Turnbull D, Fisher J (1949) J Chem Phys 17:71
82. Turnbull D (1950) J Chem Phys 18:198
83. Massa MV, Carvalho JL, Dalnoki-Veress K (2003) Eur Phys J E 12:111
84. Massa MV, Dalnoki-Veress K (2004) Phys Rev Lett 92:255509
85. Röttele A, Thurn-Albrecht T, Sommer JU, Reiter G (2003) Macromolecules 36:1257
86. Taden A, Landfester K (2003) Macromolecules 36:4037
87. Chen HL, Hsiao SC, Lin TL, Yamauchi K, Hasegawa H, Hashimoto T (2001) Macromolecules 34:671
88. Avrami M (1939) J Chem Phys 7:1103
89. Avrami M (1940) J Chem Phys 8:212
90. Avrami M (1941) J Chem Phys 9:177
91. Loo YL, Register RA, Ryan AJ (2000) Phys Rev Lett 84 18:4120
92. Chen HL, Wu JC, Lin TL, Lin JS (2001) Macromolecules 34:6936
93. Lotz B, Kovacs AJ (1969) Polym Prepr Am Chem Soc Div Polym Chem 10:820
94. Balsamo V, Urdaneta N, Pérez L, Carrizales P, Abetz V, Müller AJ (2004) Eur Polym J 40:1033
95. Gedde UW (1995) Polymer physics. Chapman and Hall, London
96. Xu JT, Fairclough JPA, Mai SM, Chaibundit C, Mingvanish M, Booth C, Ryan AJ (2003) Polymer 44:6843
97. Balsamo V, Müller AJ, Stadler R (1998) Macromolecules 31:7756
98. Balsamo V, Paolini Y, Ronca G, Müller AJ (2000) Macromol Chem Phys 201:2711
99. Müller AJ, Arnal ML, López-Carrasquero F (2002) Macromol Symp 183:199
100. Schmalz H, Abetz V, Müller AJ (2002) Macromol Symp 183:179
101. Schmalz H, Müller AJ, Abetz V (2003) Macromol Chem Phys 204:111
102. Müller AJ, Albuerne J, Esteves LM, Márquez L, Raquez JM, Degée P, Dubois P, Collins S, Hamley IW (2004) Macromol Symp 215:369
103. Müller AJ, Albuerne J, Marquez L, Raquez JM, Degée P, Dubois P, Hobbs J, Hamley IW (2004) Faraday Discuss 128:231
104. Choi YK, Bae YH, Kim SW (1988) Macromolecules 31:8766
105. Bogdanov B, Vidts AE, Chacht E, Berghmans H (1999) Macromolecules 32:726

106. Wang Y, Hillmyer M (2001) J Polym Sci Part A Polym Chem 39:2755
107. Albuerne J, Márquez L, Müller AJ, Raquez JM, Degée P, Dubois P, Castelletto V, Hamley IW (2003) Macromolecules 36:1633
108. Ho RM, Hsieh PY, Tseng WH, Lin CC, Huang BH, Lotz B (2003) Macromolecules 36:9085
109. Bhattarai N, Kim HY, Cha DI, Lee DR, Yoo DI (2003) Eur Polym J 39:1365
110. Kim JK, Park DJ, Lee MS, Ihn KJ (2001) Polymer 42:7429
111. Raquez JM, Degée P, Narayan R, Dubois Ph (2000) Macromol Rapid Commun 21:1063
112. Lendlein A, Langer L (2002) Science 296:1673
113. Shiomi T, Takeshita H, Kawaguchi H, Nagai M, Takenaka K, Miya M (2002) Macromolecules 35:8056
114. Hsieh HL (1978) J Appl Polym Sci 22:1119
115. Balsamo V, von Gyldenfeldt F, Stadler R (1996) Macromol Chem Phys 197:1159
116. Balsamo V, von Gyldenfeldt F, Stadler R (1996) Macromol Chem Phys 197:3317
117. Von Gyldenfeldt F (1999) PhD thesis, Universität Bayreuth, Germany
118. Balsamo V, Urbina de Navarro C, Gil G (2003) Macromolecules 36:4507
119. Schmalz H, Knoll A, Müller AJ, Abetz V (2002) Macromolecules 35:10004
120. Bailey TS, Pham HD, Bates FS (2001) Macromolecules 34:6994
121. Epps TH III, Bailey TS, Waletzko R, Bates FS (2003) Macromolecules 36:2873
122. Balsamo V, Gil G, Urbina de Navarro C, Hamley IW, von Gyldenfeldt F, Abetz V, Cañizales E (2003) Macromolecules 36:4515
123. DiMarzio EA, Guttman CM, Hoffman JD (1980) Macromolecules 3:1194
124. Whitmore MD, Noolandi J (1988) Macromolecules 21:1482
125. Kim G, Han CC, Libera M, Jackson CL (2001) Macromolecules 34:7336
126. Schmalz H, Böker A, Lange R, Krausch G, Abetz V (2001) Macromolecules 34:8720
127. Bailey TS, Hardy CM, Epps TH III, Bates FS (2002) Macromolecules 35:7007
128. Balsamo V, von Gyldenfeldt F, Stadler R (1999) Macromolecules 32:1226
129. Balsamo V, Stadler R (1997) Macromol Symp 117:153
130. Balsamo V, Stadler R (1999) Macromolecules 32:3994
131. Laredo E, Hernández MC, Bello A, Grimau M, Müller AJ, Balsamo V (2002) Phys Rev E 65:21807
132. Laredo E, Hernández MC, Bello A, Grimau M, Müller AJ, Balsamo V (2003) Rev Mex Fis 49:176
133. Laredo E, Hernández MC, Bello A, Grimau M, Balsamo V, Müller AJ (2003) Radiat Eff Defects Solids 158:335
134. Arnal ML, López-Carrasquero F, Laredo E, Müller AJ (2004) Eur Polym J 40:1461
135. Floudas G, Reiter G, Lambert O, Dumas P (1998) Macromolecules 31:7279
136. Floudas G, Reiter G, Lambert O, Dumas P, Yeh PH, Chu B (2000) ACS Symp Ser 739:448
137. Cohen RE, Bellare A, Drzewinski MA (1994) Macromolecules 27:2321
138. Sun G, Yang B, Jiang B (1994) Macromolecules 27:3444
139. Nojima S, Takahashi Y, Ashida T (1995) Polymer 36:2853
140. Nojima S, Yamamoto S, Ashida T (1995) Polymer J 27:673
141. Zhang R, Luo X, Ma D (1995) Polymer 36:4361
142. Drzewinski MA (1995) Macromol Symp 91:107
143. Hamley IW, Fairclough PA, Ryan AJ, Bates FS, Towns-Andrews E (1996) Polymer 19:4425
144. Gan Z, Jiang B, Zhang J (1996) J Appl Polym Sci 59:961
145. Gan Z, Jiang B, Zhang J (1997) J Appl Polym Sci 63:1793

146. Mai SM, Fairclough JPA, Viras K, Gorry PA, Hamley IW, Ryan AJ, Booth C (1997) Macromolecules 30:8392
147. Alig I, Tadjbakhsch S, Floudas G, Tsitsilianis C (1998) Macromolecules 31:6917
148. Hamley IW, Wallwork ML, Smith DA, Fairclough JPA, Ryan AJ, Mai SM, Yang YW, Booth C (1998) Polymer 39:3321
149. Nojima S, Tanaka H, Rohadi A, Sasaki S (1998) Polymer 39:1727
150. Balsamo V, Müller AJ, Gyldenfeldt FV, Stadler R (1998) Macromol Chem Phys 199:1063
151. Reiter G, Castelein G, Hoerner P, Riess G, Blumen A, Sommer JU (1999) Phys Rev Lett 83:3844
152. Nojima S, Kuroda M, Sasaki S (1997) Polym J 29:642
153. Nojima S, Kikuchi N, Rohadi A, Tanimoto S, Sasaki S (1999) Macromolecules 32:3727
154. Nojima N, Kanda Y, Sasaki S (1998) Polym J 30:628
155. Nojima S, Hashizume K, Rohadi A, Sasaki S (1997) Polymer 38:2711
156. Nojima S, Fujimoto M, Kakihira H, Sasaki S (1998) Polym J 30:968
157. Nojima S, Kakihira H, Tanimoto S, Nakatani H, Sasaki S (2000) Polymer J 32:75
158. Reiter G, Castelein G, Hoerner P, Riess G, Sommer JU, Floudas G (2000) Eur Phys J E 2:319
159. Loo YL, Register RA, Richard A, Adamson DH (2000) J Polym Sci Part B Polym Phys 38:2564
160. Zhu L, Cheng SZD, Calhoun BH, Ge Q, Quirk RP, Thomas EL, Hsiao BS, Ye F, Lotz BJ (2000) Am Chem Soc 122:5957
161. De Rosa C, Park C, Lotz B, Wittmann JC, Fetters LJ, Thomas EL (2000) Macromolecules 33:4871
162. Mai SM, Mingvanish W, Turner SC, Chaibundit C, Fairclough JPA, Heatley F, Matsen MW, Ryan AJ, Booth C (2000) Macromolecules 33:5124
163. Park C, De Rosa C, Fetters LJ, Thomas EL (2000) Macromolecules 33:7931
164. Zhu L, Huang P, Cheng SZD, Ge Q, Quirk RP, Thomas EL, Lotz B, Wittmann JC, Hsiao BS, Fengji Y, Liu L (2001) Phys Rev Lett 86:6030
165. Zhu L, Calhoun BH, Ge Q, Quirk RP, Cheng SZD, Thomas EL, Hsiao BS, Yeh F, Liu L, Lotz B (2001) Macromolecules 34:1244
166. Shiomi T, Tsukada H, Takeshita H, Takenaka K, Tezuka Y (2001) Polymer 42:4997
167. Thünemann AF, General S (2001) Macromolecules 34:6978
168. Childs MA, Matlock DD, Dorgan JR, Ohmo TR (2001) Biomacromolecules 2:526
169. Buzdugan E, Ghioca P, Stribeck N, Beckman EJ, Serban S (2001) Macromol Mater Eng 286:497
170. Fairclough JPA, Mai SM, Matsen MW, Bras W, Messe L, Turner SC, Gleeson AJ, Booth C, Hamley IW, Ryan AJ (2001) J Chem Phys 114:5425
171. Hong S, MacKnight WJ, Russell TP, Gido SP (2001) Macromolecules 34:2398
172. Hong S, MacKnight WJ, Russell TP, Gido SP (2001) Macromolecules 34:2876
173. Floudas G, Vazaiou B, Shipper F, Ulrich R, Wiesner U, Iatrou H, Hadjichristidis N (2001) Macromolecules 34:2947
174. Huang P, Zhu L, Cheng SZD, Ge Q, Quirk R, Thomas EL, Lotz B, Hsiao BS, Liu L, Yeh F (2001) Macromolecules 34:6649
175. Hong S, Yang L, MacKnight J, Gido SP (2001) Macromolecules 34:7009
176. Baiardo M, Alfonso GC (2001) Macromol Chem Phys 202:2509
177. Hong S, Bushleman AA, MacKnight WJ, Gido SP, Lohse DJ, Fetters L (2001) Polymer 42:5909
178. Reiter G, Castelein G, Sommer JU (2002) Macromol Symp 183:173
179. Xu JT, Turner SC, Fairclough PA, Mai SM, Ryan AJ (2002) Macromolecules 35:3614

180. Opitz R, Lambreva DM, De Jeu WH (2002) Macromolecules 35:6930
181. Hamley IW, Castelleto V, Floudas G, Schipper F (2002) Macromolecules 35:8839
182. Ruokolainen J, Fredrickson GH, Kramer EJ, Ryu CJ, Hahn SF, Magonov SN (2002) Macromolecules 35:9391
183. Mortensen K, Brown W, Almdal K, Alami E, Jada A (2003) Langmuir 13:3635
184. Li L, Séréro Y, Koch MH, Jeu WH (2003) Macromolecules 36:529
185. Schmalz H, Abetz V, Lange R (2003) Comp Sci Tech 63:1179
186. Zhang F, Huang H, Hu Z, Chen Y, He T (2003) Langmuir 19:10100
187. Park C, De Rosa C, Lotz B, Fetters LJ, Thomas EL (2003) Macromol Chem Phys 204:1514
188. Huang YY, Yang CH, Chen HL, Chiu FC, Lin TL, Liou W (2004) Macromolecules 37:486
189. Chen H, Li H-C, Huang YY, Chiu FC (2002) Macromolecules 35:2417
190. Avella M, Martucelli E, Greco P (1991) Polymer 32:1643
191. Xu JT, Liang GD, Fan ZQ (2004) Polymer 45:6675
192. Viras K, Mai SM, Ryan AJ, Yu GE, Booth C, Chainbundit C (2004) Macromolecules 37:3077
193. Huang P, Zhu L, Guo Y, Ge Q, Jing AJ, Chen WY, Quirk RP, Cheng SZD, Thomas EL, Lotz B, Hsiao BS, Avila-Orta CA, Sics I (2004) Macromolecules 37:3689
194. Sun L, Zhu L, Ge Q, Quirk RP, Xue C, Cheng SZD, Hsiao BS, Avila-Orta CA, Sics I, Cantino ME (2004) Polymer 45:2931
195. Xu JT, Ryan AJ, Mai SM, Yuan JJ, Cheng SY (2004) J Macromol Sci B43:685
196. Ho RM, Lin FH, Tsai CC, Lin CC, Ko BT, Hsiao BS, Sics I (2004) Macromolecules 37:5985
197. Sun L, Liu Y, Zhu L, Hsiao BS, Avila-Orta CA (2004) Polymer 45:8181
198. Nojima S, Akutsu Y, Washino A, Tanimoto S (2004) Polymer 45:7317
199. Sun J, Hong Z, Yang L, Tang Z, Chen X, Jing X (2004) Polymer 45:5969
200. Shin D, Shin K, Aamer KA, Tew GN, Russel TP, Lee JH, Jho JY (2005) Macromolecules 38:104
201. Hamley IW, Castelleto V, Castillo RV, Müller AJ, Martin CM, Pollet E, Dubois P (2005) Macromolecules 38:463
202. Jiang S, Chaoliang H, An L, Chen X, Jiang B (2004) Macromol Chem Phys 205:2229

Adv Polym Sci (2005) 190: 65–136
DOI 10.1007/12_048

Published online: 28 October 2005

Block Copolymer Micelles

Jean-François Gohy

Unité de Chimie des Matériaux Inorganiques et Organiques (CMAT)
and Centre de recherche sur les matériaux et les dispositifs électroniques micro- et nanoscopiques (CerMin), Université catholique de Louvain, Place Louis Pasteur 1, 1348 Louvain-la-Neuve, Belgium
gohy@chim.ucl.ac.be

Abstract This review summarizes recent advances to date in the area of block copolymer micelles and also tries to highlight some new directions in that field. Generalities about the preparation, characterization, and dynamics of block copolymer micelles are first outlined. Selected examples of micelle formation in aqueous and organic media are shown for block copolymers with various architectures. The different types of micellar morphologies are discussed. New directions in block copolymer micellization are finally presented.

Keywords ABC triblock copolymers · Block copolymers · Interpolyelectrolyte complexes · Micelles · Nanostructures · Polyelectrolytes · Supramolecular chemistry

Abbreviations

AFM	atomic force microscopy
AUC	analytical ultracentrifugation
b	grafting distance
BIC	block ionomer complex
CMC	critical micelle concentration
CMD	critical micelle density
CMT	critical micelle temperature
CWC	critical water concentration
DMF	*N*,*N*-dimethylformamide
DLS	dynamic light scattering
D_h	hydrodynamic diameter
FCS	fluorescence correlation spectroscopy
IPEC	interpolyelectrolyte complex
LCST	lower critical solubility temperature
MALS	multiangle light scattering
MW	molecular weight
N	degree of polymerization
N_A	degree of polymerization of core-forming block
N_B	degree of polymerization of corona-forming block
NMR	nuclear magnetic resonance
PAA	poly(acrylic acid)
PaspA	Poly(aspartic acid)
PB	polybutadiene
PBAEMA	poly(*tert*-butylaminoethyl methacrylate)
PBLG	poly(γ-benzyl-L-glutamate)
PBO	poly(butylene oxide)
PDEAEMA	poly(*N*,*N*-diethylaminoethyl methacrylate)
PDMAEMA	poly(*N*,*N*-dimethylaminoethyl methacrylate)
PDMAI	poly(*N*,*N*-dimethylaminoisoprene)
PDPAEMA	poly(*N*,*N*-diisopropylaminoethyl methacrylate)
PCEMA	poly(cinnamoylethyl methacrylate)
PCL	poly(ε-caprolactone)
PDMS	poly(dimethylsiloxane)
PEB	poly(ethylene-*co*-butylene)
PEI	poly(ethylene imine)
PEG	poly(ethylene glycol)
PEHA	poly(2-ethylhexyl acrylate)
PEHMA	poly(2-ethylhexyl methacrylate)

PEO	poly(ethylene oxide)
PEOVE	poly(2-ethoxyethyl vinyl ether)
PFS	poly(ferrocenylsilane)
PGANa	poly(sodium L-glutamate)
PHIC	poly(hexyl isocyanate)
PI	poly(isoprene)
PIB	poly(isobutylene)
PIC	poly(isocyanodipeptide)
PLL	poly(L-lysine)
PMAA	poly(methacrylic acid)
PMANa	poly(sodium methacrylate)
PMACs	poly(cesium methacrylate)
PMEMA	poly(2-(*N*-morpholinoethyl) methacrylate)
PMMA	poly(methylmethacrylate)
PMOVE	poly(2-methoxyethyl vinyl ether)
PMOXA	poly(methyl oxazoline)
PMPS	poly(methylphenylsilane)
PMTGEVE	poly(methyl tri(ethyleneglycol) vinyl ether)
PMVE	poly(methyl vinyl ether)
PNIPAAM	poly(*N*-isopropyl acrylamide)
PPO	poly(propylene oxide)
PPQ	poly(phenylquinoline)
PS	polystyrene
PSCNa	poly(sodium 4-styrene carboxylate)
PSMA	poly(stearylmethacrylate)
PSOH	poly(styrene-*co*-[*p*-(2,2,2-trifluoro-1-hydroxy-1-trifluoromethyl)ethyl-methylstyrene])
PSS	poly(styrene sulfonate)
PTAN	poly(1,1,2,2-tetrahydroperfluorooctyl acrylate)
PtBA	poly(*tert*-butylacrylate)
PtBMA	poly(*tert*-butyl methacrylate)
PtBS	poly(*tert*-butylstyrene)
PVA	poly(vinyl alcohol)
PVAC	poly(vinylacetate)
P2VP	poly(2-vinylpyridine)
P2VPQ	poly(2-vinylpyridine) quaternized
P4VP	poly(4-vinylpyridine)
R_c	micellar core radius
R_g	radius of gyration
R_h	hydrodynamic radius
R_m	total micellar radius
SANS	small-angle neutron scattering
SAXS	small-angle X-ray scattering
SEC	size-exclusion chromatography
SEM	scanning electron microscopy
$scCO_2$	supercritical CO_2
SLS	static light scattering
TEM	transmission electron microscopy
T_g	glass transition temperature
THF	tetrahydrofuran

UCST upper critical solubility temperature
Z aggregation number

1 Introduction

Block copolymers have been the focus of much interest during the last 30 years because their constituent blocks are generally immiscible, leading to a microphase separation. Since the different blocks are linked together by covalent bonds, the microphase separation is spatially limited and results in self-assembled structures whose characteristic sizes are of the order of a few times the radius of gyration, R_g, of the constituent blocks and thus range from ca. 10 to 100 nm [1].

Moreover, these self-assembled structures tend to be regularly distributed troughout the bulk material, giving rise to long-range ordering and the formation of structures such as cubic array of spheres or hexagonally packed cylinders. However, the order in these soft materials is not as perfect as in crystals, and the size of the ordered domains or grains is generally limited to a few micrometers or less. Several annealing methodologies in which block copolymer materials have been exposed to a suitable solvent, temperature, or electric field have been therefore developed to improve the ordering.

Predicting the characteristic sizes and morphologies of these nanostructures has been an intense topic of investigation from both the theoretical and experimental points of view. Critical parameters are the degree of polymerization and the volume fraction of the constituent blocks, as well as the Flory–Huggins parameter between them. More complete information about microphase separated structures in bulk block copolymers can be found in the book of Hamley [2].

Whenever block copolymers are dissolved in a selective solvent that is a thermodynamical good solvent for one block and a precipitant for the other, the copolymer chains may associate reversibly to form micellar aggregates in deep analogy with the situation observed for classical low-MW surfactants. In this respect, a CMC can be defined and experimentally measured for block copolymer micelles, as discussed in Sect. 2.1. Compared to low-MW surfactants, the values of the CMC are much lower in the case of block copolymer macrosurfactants. This motivates, e.g., the use of block copolymer micelles as nanocontainers for drug delivery. In contrast to low-MW surfactants, these block copolymer nanocontainers do not dissociate into unimers whenever they are diluted in the blood stream and can therefore transport the drugs to a specifically targeted area provided that they are functionalized by suitable moeities for site-recognition [3]. Nevertheless, macromolecular chains can encounter some dissolution problems whenever they are placed in a selective solvent. This problem is especially emphasized for block copolymers containing a high T_g or a large insoluble block. A way to improve solubility is based on the temporary use of an organic solvent, which then allows the formation of micelles. Therefore, equilibrium micelles have to be distinguished for nonequilibrium ones, as will be discussed in Sect. 2.2.

From a morphological point of view, block copolymer micelles consist of a more or less swollen core resulting from the aggregation of the insoluble blocks surrounded by a corona formed by the soluble blocks, as decribed in Sect. 2.3. Experimental techniques that allow the visualization of the different compartments of block copolymer micelles will be presented in Sect. 2.4. Other techniques allowing micellar MW determination will also be briefly discussed. Micellar dynamics and locking of micellar structures by cross-linking will be commented on in Sects. 2.5 and 2.6, respectively.

Selected examples of block copolymer micelles in both aqueous and organic media will then be presented in Sects. 3 and 4. Section 4.3 emphasizes stimulus-responsive micellar systems from double-hydrophilic block copolymers. Prediction of the dimensional characteristic features of block copolymer micelles and how it varies with the composition of the copolymers will be shortly outlined in Sect. 5, with a consideration of both the theoretical and experimental approaches. Tuning of micellar morphology and triggering transitions between different morphologies will then be discussed in Sect. 6.

Recent progress in novel micellar structures, including micelles containing "exotic" blocks such as natural or synthetic polypeptides and metal-containing segments, micelles from ABC triblock copolymers, Janus micelles and other noncentrosymmetric micelles, micelles based on interpolyelectrolyte or other noncovalent complexes, and metallosupramolecular micelles, will be discussed in Sect. 7.

With the increasing number of publications on block copolymer micelles (a database literature search with these three associated keywords already gives more than 500 references), an exhaustive description of all previous works would not be possible in the framework of the present review. This contribution has rather as its purpose giving a general overview about block copolymer micelles for the nonspecialist and will therefore try to answer such practical questions as how does one prepare block copolymer micelles? How does one characterize them? What are the different types of structures that can be formed? How can we predict them? How does one tune the morphology of these micelles? These basic questions and the corresponding answers will be illustrated by selected examples. Then, we will focus on the new directions that are currently implemented in this field.

The research area of block copolymer micelles has been reviewed by other authors including Price [4], Piirma [5], Tuzar and Kratochvil [6], Riess and coworkers [7, 8], Webber et al. [9], Alexandridis and Hatton [10], Nace [11], Hamley [2], Alexandridis and Lindman [12], and Xie and Xie [13]. A very complete review on block copolymer micelles was published recently by

Riess [14], while polyelectrolyte block micelles have been surveyed by Förster et al. [15].

In the following discussion, block copolymers will be simply designated by the acronym A-B for a diblock copolymer, A-B-A for a triblock copolymer with two identical outer blocks, A-B-C for an ABC triblock copolymer, etc. A complete list of abbreviations for the A, B, and C polymer blocks is given in the Abbreviations and Symbols section.

2 Block Copolymer Micelles: General Features

Generalities about block copolymer micelles have been reviewed by Hamley [2] and Riess [14], based on previous works from the 1980s and 1990s. This topic will not be covered in detail, but the basic principle, as well as some important practical issues, will be reviewed. The essential experimental techniques used for block copolymer micelle characterization will also be outlined briefly.

2.1 Critical Micelle Concentration

Whenever amphiphilic block copolymer chains are dissolved at a fixed temperature and in a selective solvent for one of the blocks, they self-associate through a closed association process to form micelles similarly to low-MW surfactants.

The critical concentration at which the first micelle forms is called the critical micelle concentration, or CMC. As the concentration of block copolymer chains increases in the solution, more micelles are formed while the concentration of nonassociated chains, called unimers, remains constant and is equal to the value of the CMC. This ideal situation corresponds to a system at thermodynamic equilibrium. However, experimental investigations on the CMC have revealed that its value depends on the method used for its determination. Therefore, it seems more reasonable to define phenomenologically the CMC as the concentration at which a sufficient number of micelles is formed to be detected by a given method [16]. In practical terms, the CMC is often determined from plots of the surface tension as a function of the logarithm of the concentration. The CMC is then defined as the concentration at which the surface tension stops decreasing and reaches a plateau value.

Information about the process associated with the CMC has been provided by Tsunashima et al. [17], who investigated the micellization of a PS-PB diblock in various solvents. In these studies, DLS was used to measure R_h. Measurements were performed in solvents having the same refractive index of either the PS or PB chains, allowing the determination of the characteristic sizes of the individual blocks. In nonselective solvents, it was confirmed that the PS-PB chains were molecularly dissolved and that both blocks adopted a stretched conformation due to intersegmental repulsive interactions. In *n*-decane, a selective solvent for the PB block, PS-PB unimers with collapsed PS segments were observed at low concentration while aggregated PS-PB chains forming spherical micelles were observed at higher concentrations, thus defining the CMC.

How the CMC varies with the composition of the copolymer has been extensively studied for PEO-PPO-PEO triblock copolymers in aqueous solution [18–24]. It was found that copolymers with a larger hydrophobic PPO block had a lower CMC. The CMC was indeed found to decrease exponentially with the PPO block length [18–20] while an increase in the number of EO units only resulted in a small increase of the CMC [19]. At a constant PPO/PEO ratio, the CMC was found to decrease with increasing total MW of the copolymer [19].

A critical micelle temperature or CMT is a very useful value for PEO-PPO-PEO copolymers. This arises from the fact that micellization in these copolymers is due to the dehydration of the PPO block with increasing temperature. The value of the CMT ranges from 20 to 50 °C in commercially available PEO-PPO-PEO copolymers. The CMT increases whenever the copolymer concentration is increased [19].

The influence of the copolymer chain architecture on the CMC has been investigated in PEO- and PBO-containing bock copolymers, as schematically depicted in Fig. 1. From an entropic point of view, the formation of micelles should be favored for diblock architectures compared to triblock and cyclic ones, the reason being that two block junctions should reside at the

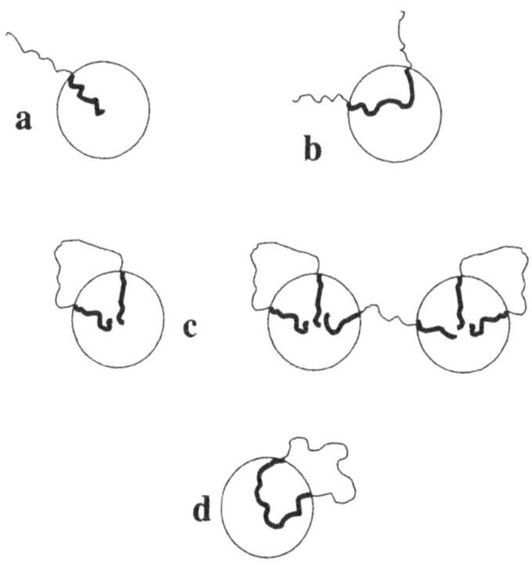

Fig. 1 Schematic representation of chain conformation in micelles from **a** linear PEO-PBO diblock copolymers, **b** linear PEO-PBO-PEO triblock copolymers, **c** linear PBO-PEO-PBO triblock copolymers and **d** cyclic PEO-PBO diblock copolymers

core–corona interface for triblock and cyclic copolymers, while unfavorable loop formation will not be observed for the diblock copolymer. Finally, the less entropically favored situation will be observed for the cyclic copolymers since both of their blocks should form a loop. Experimental investigations by Booth et al. on PEO_{2m}-PBO_{2n} and PBO_n-PEO_{2m}-PBO_n [21] revealed that the CMC is ca. two orders of magnitude lower for the diblocks at a constant n value. The CMC for PEO_m-PBO_{2n}-PEO_m was found to be intermediate between the two latter architectures [21]. Surprisingly enough, micellization was favored for the cyclic copolymer compared to the triblock [21]. This can be understood by considering that the conformation of the cyclic copolymer is restricted in both the unassociated and associated states and therefore has no impact on the Gibbs free energy of micellization. In addition to the above-mentioned considerations, bridging of chains between micelles could occur in the case of PBO-PEO-PBO triblock copolymers, as confirmed by DLS measurements [25]. In very dilute solutions, flowerlike micelles with all the triblock chains making loops are likely formed while bridging and formation of larger structures were already detected for a concentration of 1 wt% as evidenced by a markedly increased viscosity of the solution [25].

The effect of the diblock versus starlike architecture on the CMC of PIB-PMVE copolymers was recently investigated in water by Faust and coworkers [26]. The CMC of an AB (A standing for PIB and B for PMVE) was compared to the values obtained for the corresponding $2A_1B$ and A_2A_3B copolymers, where $N(A) = 2N(A_1) = N(A_2) + N(A_3)$, and $N(B)$ is constant. Previous work of Pispas et al. on PI-PS miktoarm copolymers dissolved in n-decane revealed that the micellar characteristic features such as Z and R_h were strongly dependent on the architecture, but the corresponding CMCs could not be determined [27]. Faust et al. followed the fluoresence of a probe (pyrene) in order to determine the CMC. In this widely used technique, the flurorescence intensity ratios of two different bands ($I_{338.3}/I_{335.9}$) of pyrene is measured as a function of copolymer concentration. An increase in this ratio is observed at a given copolymer concentration, indicating the onset of micelle formation and thus the CMC. The CMC was found to increase in the order $2A_1B$, A_2A_3B, AB. Although the absolute values of the CMCs measured by other techniques (DLS and surface tension measurements) were significantly different from those obtained by pyrene fluorescence, the same order in CMCs was always observed. It can then be concluded that the shielding of a hydrophobic A chain by a hydrophilic B chain is less efficient when A is branched. According to Faust's report, a branched hydrophobic A chain therefore displays properties of a linear chain of higher MW [26].

For highly asymmetric block copolymers with a large insoluble block, the copolymer chains can't be directly solubilized in the selective solvent. However, micelles can be obtained from these copolymers by the temporary use of a nonselective solvent which is further eliminated (see Sect. 2.2 for further details on this issue). In principle, all the copolymer chains are aggregated for these systems and the CMC is therefore estimated to be infinitely low, as discussed by Eisenberg et al. [28].

At the end of this section, we will comment on the phenomenon of anomalous micellization that has been reported by several groups on various block copolymer micelles [6, 19, 29, 30]. The anomalous micellization behavior is characterized by the observation of large particles prior to the onset of micellization. This is macroscopically visualized by an opalescence of the solution. Chu and Zhou investigated by DLS this process and found that large particles were coexisting with either unimers or unimers and micelles [19]. These large particles represented a few percent of the total mass. It was demonstrated that the anomalous micellization resulted from composition polydispersity of the starting copolymer. Even copolymers with a narrow size polydispersity index could contain small amounts of copolymers with much larger insoluble blocks that perturb the micellization behavior. This hypothesis was credited by elimination of the initial large particles, leading then to solutions with a "normal" micellization behavior [31].

2.2 Preparation of Block Copolymer Micelles

Although it strongly influences the micellar characteristic features, the method used for the preparation of block copolymer micelles has been very often poorly discussed in the literature. This crucial point was raised in the excellent review of Riess [14].

The most direct way to prepare a block copolymer micellar solution consists in the direct dissolution of the bulk sample in a selective solvent for one of the blocks. However, this method generally only works if the total MW of the copolymer is low and the length of the insoluble block is short enough. A way to improve solubility, then, consists in "annealing" the solution by prolongated stirring, thermal, or ultrasound treatments. According to the literature, these techniques have the disadvantage that they lead, depending on the block copolymer, to nonequilibrium micelles, especially when the core-forming chains of the block copolymer are below T_g [32]. The characteristic features of the resulting micelles will then depend on the bulk morphology of the starting bulk sample. It can be assumed that the annealing treatment will result rather in the dispersion of bulk particles into the selective solvent than in the formation of micelles based on a unimer–aggregate equilibrium. These particles will be protected against flocculation by the solvated chains at their outer surface, but their morphology will be essentially dictated by the annealing conditions [32].

It should, however, be mentioned that the transfer of a bulk-organized system into solution can lead to very interesting structures, as will be demonstrated in Sect. 7.3 in the case of Janus micelles [33]. In this case, a micellar structure is preformed in the bulk, its core is stabilized by cross-linking, and the whole material is finally dissolved in a very good solvent for the coronal chains. A similar stabilization of a bulk structure that was then transferred to solution was implemented by Liu et al. for PCEMA-containing block copolymers [34].

Another method is based on the dissolution of the block copolymer in a nonselective solvent, resulting in the formation of molecularly dissolved chains. The properties of this solvent are further changed in order to trigger aggregation of the dissolved chains and hence micelle formation. A selective solvent for one of the blocks and precipitant for the others is generally added to the molecularly dissolved chains, although other ways such as temperature or pH changes can be used for micellization as well. In this case, the unimer–micelle equilibrium can be generally reversibly tuned by pH or temperature changes. These systems are then rather considered as stimulus-responsive micelles that can be disassembled or assembled depending on the applied stimulus, as will be discussed in Sect. 4.3.

Addition of a selective solvent to molecularly dissolved chains has been used by many research teams to prepare block copolymer micelles. The initial nonselective solvent can be further eliminated by evaporation or can be gradually replaced by the selective solvent via a dialysis process. The stepwise dialysis initially introduced by Tuzar and Kratochvil is now widely used for micelle preparation [6], especially for the formation of aqueous micelles [32].

This technique does not, however, overcome the formation of "frozen" micelles due to the formation of glassy cores at a specific nonselective solvent/selective solvent composition. Polydisperse micelles can also be generated during this preparation process if the starting material is characterized by a composition or MW polydispersity. In this respect, micelles will be first formed by the chains containing the larger insoluble block during the addition of the selective solvent.

This method offers, nevertheless, several advantages. Firstly, the formation of large aggregates can be suppressed for block copolymers that were previously solubilized directly in the selective solvent. Secondly, it allows the formation of micelles for highly asymmetric copolymers with a large insoluble block, as illustrated by the works of Eisenberg et al. on the so-called "crew-cut" micelles [28, 35, 36]. "Crew-cut" micelles are prepared from highly asymmetric diblock copolymers containing very short water-soluble blocks initially dissolved in a nonselective solvent (most often DMF). Water is then slowly added to these solutions. After addition of a critical amount of water defined as the CWC, aggregation is observed. An additional amount of water is added in order to freeze in the morphology of the formed aggregates generally characterized by a high T_g PS core. Observations on the CWC have been realized by turbidimetry and rationalized by Eisenberg et al. [37]. These authors have shown that the CWC depends on both the copolymer concentration in nonselective solvent and the MW of the insoluble block. The higher the copolymer concentration and the MW of the insoluble block, the lower the CWC [37]. The same authors also investigated the influence of the starting nonselective solvent on the characteristic features of the resulting "crew-cut" micelles and found that polymer-solvent interactions determined the dimensions of both the core and corona of the aggregates [38].

2.3 Micellar Structure

It is important to define clearly the characteristic features of block copolymer micelles. We mentioned above that the insoluble blocks formed a micellar core surrounded by a corona. Depending on the composition of the starting block copolymer, two limiting structures can be drawn: (1) "starlike" micelles with a small core compared to the corona and (2) crew-cut micelles with a large core and highly stretched coronal chains. Both situations are schematically depicted in Fig. 2.

The process of micellization is thus characterized by the aggregation of a given number of block copolymer chains, defined as the aggregation number or Z. The core is characterized by its radius R_c (Fig. 2), while the overall radius of the micelle is defined as R_m. Other ways of defining the overall dimension of a micelle are the radius of gyration R_g and the hydrodynamic radius R_h; both are defined elsewhere [39]. The distance between neighboring blocks at the core/corona interface is called the grafting distance b. Therefore, b^2 is the area occupied by one chain at the core/corona interface and can be compared to the area by head group as defined for low-MW surfactant micelles [40].

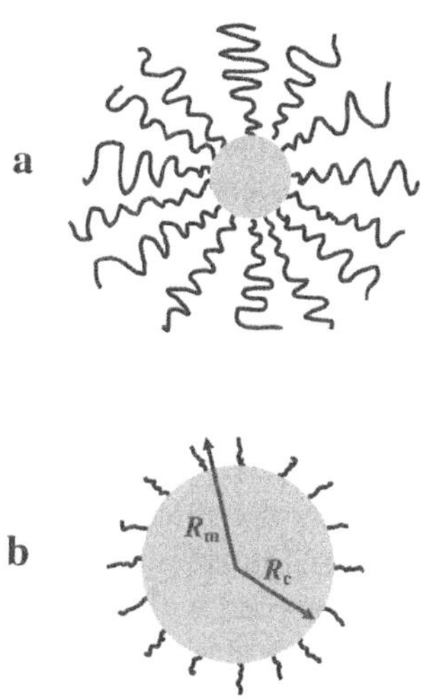

Fig. 2 Schematic representation of a "starlike" (**a**) and a "crew-cut" (**b**) micelle. Important structural parameters (R_c and R_m) of block copolymer micelles are indicated in (**b**)

Key parameters that control R_c, R_m, b, and Z are the degree of polymerization of the polymer blocks, N_A and N_B, and the Flory–Huggins interaction parameter χ. The free energy of a micelle is mainly determined by (1) the interfacial energy of the core/shell interface, (2) the stretching energy of the block copolymer chains, and (3) the repulsion among coronal chains. The minimum of the free energy corresponds to an equilibrium grafting distance b, which depends on block lengths and salt concentration in the case of charged coronal blocks.

The micellar structure depicted in Fig. 2 is of course only valid for simple AB diblock copolymers. The situation can be much more complex for micelles prepared from block copolymers with complex architectures, as will be discussed later.

2.4 Characterization of Block Copolymer Micelles – Experimental Techniques

Extensive reviews on experimental techniques suitable for block copolymer micelle characterization have been provided by Tuzar [41], Munk [42], Chu and Zhou [19] Webber [43], Mortensen [44], Zana [45], and Hamley [2]. Moreover, Hamley has systematically listed the different techniques specifically used for different types of block copolymer micelles.

In the following discussion, we would like to briefly highlight the different methods used for morphological characterization of block copolymer micelles emphasizing advantages and limitations. This will be illustrated by selected examples from our own investigations on various block copolymer micelles.

Transmission electron microscopy (TEM) techniques have been widely used over the last 30 years for the direct visualization of block copolymer micelles. Different ways of sample preparation have been tested, some of them being translated from TEM observations on soft biological materials. Since block copolymer micelles essentially contain light elements such as C, H, N, and O, no good electronic contrast can be obtained from these specimens. Moreover, these soft materials are subjected to degradation during observation, a "ghost" picture of the objects being then recorded. Most of the preparation techniques have been devoted to enhancing both the electronic contrast and the stability of the micellar objects for TEM observations. Contrasting techniques with heavy metals have been widely used to reach these two goals. In these experiments the micelles are usually initially deposited on a TEM grid covered with an amorphous film and further observed in the dried state. Heavy metals can interact with block copolymer micelles bearing suitable functional moieties. Typical examples are the reactions of RuO_4 to aromatic double bonds and OsO_4 to aliphatic double bonds. These staining techniques also allow the direct visualization of a specific compartment of the micelles, e.g., the core. This is illustrated in Fig. 3 for aqueous micelles from a PS-P2VP-PEO triblock copolymer and contrasted with RuO_4 [46, 47]. At basic pH, both the PS core and hydrophobic P2VP shell are equally stained and cannot be distinguished, while at acidic pH the protonated and water-soluble P2VP shell is less contrasted and can be discriminated from the PS core. The PEO corona is never visualized in these experiments.

Negative contrasting techniques are based on the use of a contrasting agent that does not interact with the micelles and that therefore spreads onto the substrate. Micelles then appear as bright spots on a dark background. Phosphotungstic acid is a typical negative staining agent. The dimension of the core + dried corona is measured from this experiment. Metal can be splattered on block copolymer micelles by exposing the TEM grid to metal vapor

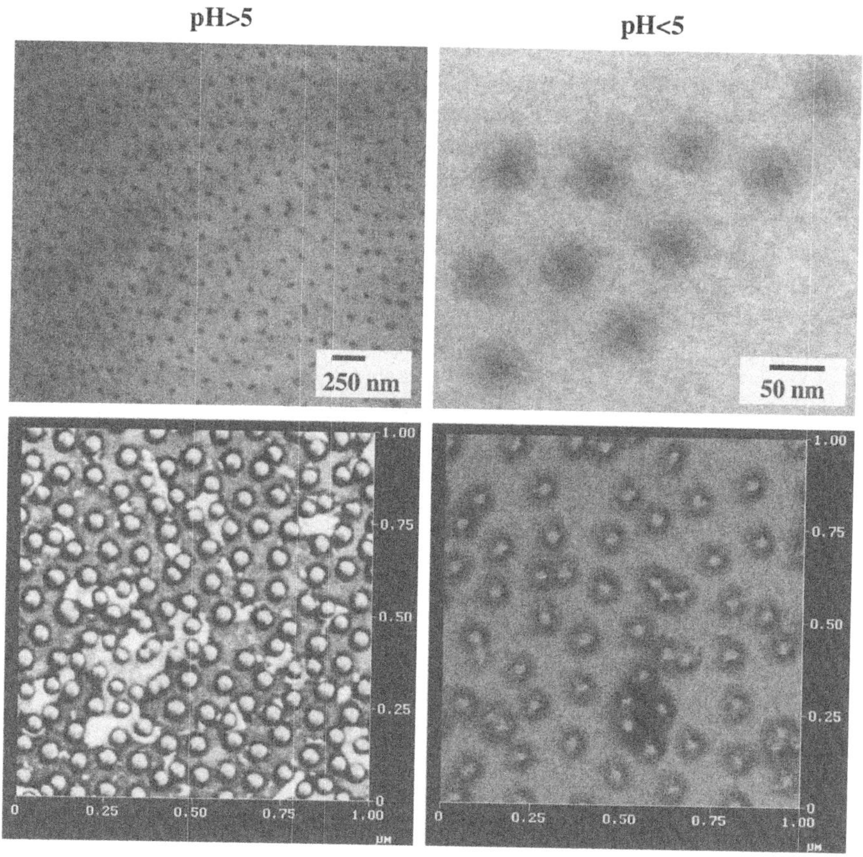

Fig. 3 TEM (*top*) and AFM phase contrast images (*bottom*) of aqueous micelles formed by a PS_{200}-$P2VP_{140}$-PEO_{590} ABC triblock copolymer at $pH > 5$ (*left*) and $pH < 5$ (*right*). For TEM pictures, the PS and P2VP blocks have been stained by RuO_4. AFM images have been recorded with tapping mode (contrast scale: *black*: 0°, *white*: 45°). Adapted from [47]

under a certain angle. Each micelle will generate a shadow whose size and shape depend on the metal deposition angle and on the morphological characteristic features of the micelles. A typical TEM picture from Pt-shadowed PS-PEO metallosupramolecular micelles (see Sect. 7.5 for further details) showing large aggregates is presented in Fig. 4 [48]. Metallic replica of freeze-fractured micellar solution can also be examined.

The more recently developed cryo-TEM technique has started to be used with increasing frequency for block copolymer micelle characterization in aqueous solution, as illustrated by the reports of Esselink and coworkers [49], Lam et al. [50], and Talmon et al. [51]. It has the advantage that it allows for direct observation of micelles in a glassy water phase and accordingly determines the characteristic dimensions of both the core and swollen corona provided that a sufficient electronic contrast is observed between these two domains. Very recent studies on core-shell structure in block copolymer micelles as visualized by the cryo-TEM technique have been reported by Talmon et al. [52] and Förster and coworkers [53]. In a very recent investigation, cryo-TEM was used to characterize aqueous micelles from metallosupramolecular copolymers (see Sect. 7.5 for further details) containing PS and PEO blocks. The results were compared to the covalent PS-PEO counterpart [54]. Figure 5 shows a typical cryo-TEM picture of both types of micelles.

Scanning electron microscopy (SEM) techniques have proven to be suitable for the visualization of block copolymer micelles, as illustrated in, e.g., the recent work of Erhardt et al. on "Janus" micelles (Sect. 7.3) [55].

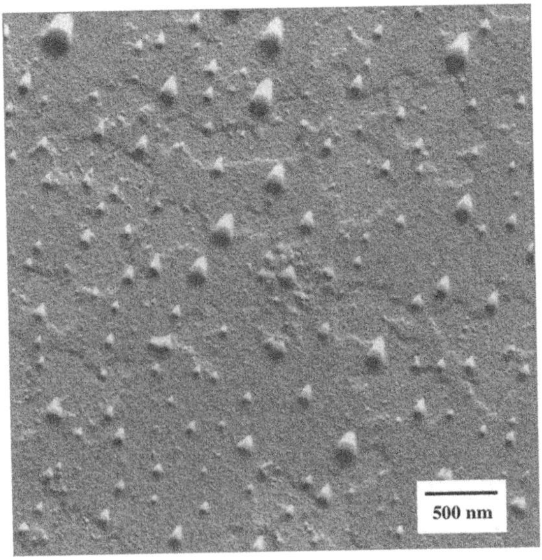

Fig. 4 Pt-shadowed TEM picture of micelles and aggregates formed by metallo-supramolecular PS_{20}-[Ru]-PEO_{70} aqueous micelles. Reprinted with permission from [48]. Copyright (2003) Wiley

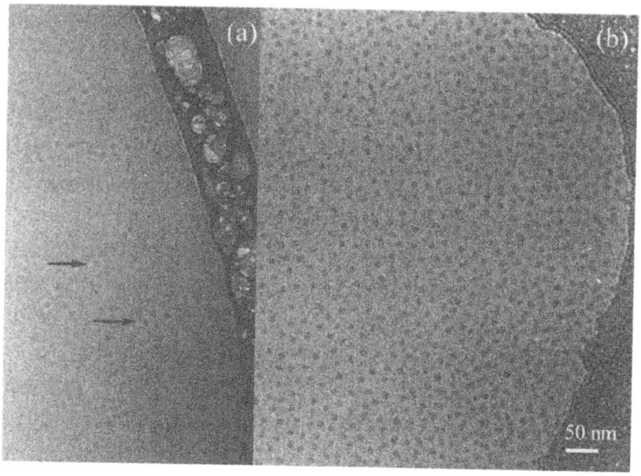

Fig. 5 Cryo-TEM pictures of metallosupramolecular PS_{20}-[Ru]-PEO_{70} (**a**) and covalent PS_{22}-PEO_{70} (**b**) aqueous micelles. *Arrows* in **a** indicate isolated micelle and cluster of micelles. Reprinted with permission from [54]. Copyright (2004) Springer

Direct visualization of block copolymer micelles can also be achieved by atomic force microscopy (AFM). AFM measurements can be performed either in the dried state or directly "in situ" within a liquid cell. "Dry" measurements are realized on micelles adsorbed or deposited on a flat solid substrate (silicon wafer, mica, ...) that have been further dried. The shape and size of the micelles can however be affected by tip-convolution effects, by specific interactions between the substrate and some moeities of the block copolymer, or by relaxation of a low T_g micellar core resulting in the flattening of the micelles on the substrate. Typical AFM pictures from dried PS-P2VP-PEO aqueous micelles at basic or acidic pH are shown in Fig. 3 and can be directly compared with the corresponding TEM pictures [46, 47]. At first glance, AFM measurements in a liquid cell seem to be more appropriate for obtaining pertinent morphological information on micelles in their original swollen state. In these AFM experiments, the micelles have a more or less developed tendency to adsorb at the substrate/liquid interface, which is the locus where the measurements are performed. Nevertheless, several artifacts, such as sticking of the micelles on the AFM tips, can perturb the observation as highlighted in very recent papers of Hamley et al. [56, 57].

Information on micellar morphology can also be obtained by scattering techniques with the advantage of giving a mean value calculated over a large number of micelles, but with the drawback that it is model dependent. A recent review on small-angle X-ray scattering on block copolymer micelles was published by Mortensen [44]. Small-angle neutron scattering (SANS) has also been widely used to elucidate micellar structures. In order to generate a sufficient scattering contrast, SANS experiments are generally conducted in deuterated solvents and/or with partially deuterated copolymers. Under such labeling conditions, the core and corona of the micelles can be selectively studied. Light scattering is a very common technique used for block copolymer micelle characterization. Static light scattering allows one to determine the absolute weight-averaged MW and thus Z of the micelles, as well as R_g. Information about the quality of the selective solvent for the coronal chains through the second virial coefficient can also be obtained [39]. R_h can be measured by dynamic light scattering [39].

Analytical ultracentrifugation (AUC) has been used infrequently to characterize block copolymer micelles, although it allows one to determine the weight fraction of micelles, unimers, and eventually other species such as superaggregates of micelles. The sedimentation coefficient of the different species in solution depends on their MW, buoyancy, and friction coefficient, and on the centrifugal force applied. Although it can be argued that the unimer-micelle-aggregate equilibrium is continuously disturbed and reestablished during AUC experiments, it has been successfully applied to block copolymer micelles as demonstrated by the early works of Selb and Gallot [58] and the recent investigations of Schubert et al. [59, 60]. In this latter work, the different species present in solution was fractionated using a sucrose gradient, and the MW of each fraction was also measured.

Size-exclusion chromatography (SEC) has been used to characterize the unimer-micelle distribution. However, SEC is not an absolute method and thus requires calibration. Since it is practically impossible to calibrate a SEC apparatus for the unimers and micelles formed by a block copolymer, only indicative MW values can be obtained. Moreover, several authors have noted a strong perturbation of the unimer-micelle equilibrium during SEC experiments even when interaction of the material with the SEC column was minimized [4, 61, 62].

Other characterization techniques are devoted to the determination of the CMC. They have been reviewed elsewhere [2, 14] and will not be discussed here. Various NMR, fluorescence, and stop-flow techniques have also been used to characterize different aspects related to the dynamics of block copolymer micelles, as will be discussed in the next section.

2.5 Dynamics of Block Copolymer Micelles

The dynamics of micellar systems is a very important concern that is relevant at several levels. In this respect, the dynamics required for the establishment of the unimer–micelle equilibrium is one aspect. The so-called problem of micelle hybridization that deals with the exchange rate of unimers between different micelles is closely related. Finally, the chain dynamics of polymer blocks in either the core or the corona is another concern that can, however, be linked to some extent to the first two mentioned concepts. These different aspects have been scarcely studied and reviewed by Tuzar and Kratochvil [6, 41], and

Hamley [2] and recently by Riess [14]. However, we feel that these important problems should be briefly discussed in the present review.

Stop-flow experiments have been performed in order to study the kinetics of micellization, as illustrated by the work of Tuzar and coworkers on PS-PB diblocks and the parent PS-PB-PS triblocks [63]. In these experiments, the block copolymers are initially dissolved as unimers in a nonselective mixed solvent. The composition of the mixed solvent is then changed in order to trigger micellization, and the scattered light intensity is recorded as a function of time. The experiment is repeated in the reverse order, i.e., starting from the block copolymer micelles that are then disassembled by a change in the mixed solvent composition. The analysis of the experimental results revealed two distinct processes assigned as unimer–micelle equilibration at constant micelle concentration (fast process) and association–dissociation equilibration, accompanied by changes in micellar concentration (slow process).

The differences observed between AB di- and ABA triblock copolymers could be explained because two A blocks must escape from the micellar core in the case of ABA triblock chains.

The exchange of block copolymer chains between micelles is also a problem of high interest. It is especially relevant whenever applications for block copolymer micelles are considered. For example, block copolymer micelles can be used to stabilize colloids such as pigment particles [64, 65]. In this case, it is generally acknowledged that some block copolymer chains escape from the micelles and are then adsorbed on the surface of the colloidal particle or that the whole micelle is adsorbed and then rearranges at the surface of the particle [14]. For such processes to be effective, the block copolymer chains should contain a good anchoring block to interact with the particle, the other block should dangle in solution and ensure a good stabilization of the particle, and finally the block copolymer chains should escape easily from the micelles or the micelles should easily rearrange whenever they are in contact with the particle. The unimer–micelle exchange rate has been studied by Creutz et al. using steady-state fluorescence [66, 67]. In such experiments, micelles from PDMAEMA-PMANa copolymers labeled with a covalently bound donor, naphthalene, were mixed with micelles formed from the same copolymers in which the donor had been replaced by an acceptor, pyrene (pyrene was not bound to the copolymer but simply mixed). Upon mixing the two solutions, exchange of PDMAEMA-PMANa chains occurred and donor-tagged chains entered micelles containing the acceptor. Specifically exciting the donor and monitoring the emission of the acceptor has been simultaneously performed in order to measure the energy transfer between the two molecules. The increasing emission intensity of the acceptor with time was then considered as a measure of the exchange rate of unimers between micelles. The rate constants for the PDMAEMA-PMANa copolymers were found to be of the order of 10^{-3} s^{-1}. Moreover, the influence of the molecular architecture and the composition of the copolymers was also studied. The exchange rate was slower whenever the length of the insoluble block was increased and a triblock architecture with the outer insoluble blocks was considered. The influence of temperature was studied by the same authors on aqueous micelles formed by PS-PMANa and PtBS-PMANa copolymers. No exchange rate could be detected at room temperature, while the exchange rate was measurable at 60 °C. The influence of the addition of a small amount of a good solvent for the micellar core was also studied as a way to increase the exchange rate in frozen micelles. In a similar way, the addition of a plasticizer such as dimethyladipate has been successfully used to increase the micelle–unimer exchange rate in frozen PMMA-PMANa micelles [68].

Similar fluorescence techniques have been used by other authors on micelles containing PS cores. In this respect, Riess and Hurtrez showed that no chains were exchanged for PS-PEO micelles, even for rather low PS MW [69].

The chain-exchange in pyrene and naphthalene singly labeled PEO-PMAA micelles was recently investigated by fluorescence spectroscopy [70]. The driving force for micelle formation in these copolymers is the self-complexation between PEO and PMAA blocks due to hydrogen bonding. The chain-exchange between these micelles at a low degree of ionization of the PMAA blocks was observed to take place via two distinct mechanisms. Insertion and expulsion of the single chains were found to be responsible for a faster exchange event, whereas merging and subsequent splitting of the micelles was characteristic for the slower exchange process.

SANS has been recently used to study problems related to micelle preparation and kinetics, as reported by Bates and coworkers who have used time-resolved SANS to study molecular exchange and micelle equilibration for PEO-PB diblocks in water [71]. The authors have shown that the micellar structures initially formed upon dissolution were completely locked in up to 8 d after preparation. Fluorometry and DLS have also been used to monitor micelle equilibration [72].

Exchange of unimers between two different types of block copolymer micelles has often been referred to as hybridization. This situation is more complex than for the case described above because thermodynamic parameters now come into play in addition to the kinetic ones. A typical example of such hybridization is related to the mixing of micelles formed by two different copolymers of the same chemical nature but with different composition and/or length for the constituent blocks. Tuzar et al. [41] studied the mixing of PS-PMAA micelles with different sizes in water–dioxane mixtures by sedimentation velocity measurements. These authors concluded that the different chains were mixing with time, the driving force being to reach the maximum entropy.

The last feature about micellar dynamics is related to the local mobility of chain segments in the core or in the corona of the micelles. SAXS, SANS, and fluorescence techniques have proven to be effective for obtaining information about chain conformation and dynamics in the different micellar domains, as reviewed by Alexandridis and Hatton [10], Prochazka et al. [73], and Zana [45]. Neutron spin echo has been used to measure the dynamic structure factor from intensity fluctuations as a function of time. Farago et al. have used this technique to probe the dynamics of the coronal PI chains for PS-PI micelles in decane [74]. The results were found to be in agreement with the model of de Gennes for "breathing" coronal chains. A recent investigation of Castelletto et al. by neutron spin echo dynamics on PBO-PEO micelles in D_2O revealed the presence of two dynamic modes, i.e., "fast" and "slow" modes, whatever the concentration of the solution [75]. The "slow" mode that dominates at low scattering angles was shown to correspond to the translational diffusion of micelles, while the additional "fast" mode was ascribed to internal "blob scattering" in the micellar corona.

Direct information on chain mobility can be easily obtained by NMR while considering that a decreased mobility of protons in polymer chains with hindered motion results in broadening of the respective NMR lines and even disappearance of the corresponding signal when the polymer is in the glassy state. This last feature has been widely used to prove micelle formation as illustrated by Spevacek [76] on PS-PB micelles with a PS core. At room temperature, the aromatic signals associated to PS chains could not be detected while they were visible at 87 °C. The same approach was more recently used to study stimulus-responsive micelles that could change their structure depending on the environment. This is illustrated by the work of Armes et al. on so-called schizophrenic micelles [77]. Depending on pH and/or temperature, core-forming chains can be switched to coronal chains while the reverse is observed for the initial coronal chains that transform into core chains. This reorganization of the micellar structure upon application of the stimulus can be easily followed by NMR.

^2H NMR measurements on partially deuterated PS-PMANa copolymers have been performed by Gao et al. [78]. Deuterated segments were introduced in several locations of the core-forming chains, and their mobility was measured. For segments buried in the PS micellar core, the mobility was found to be essentially the same as in a bulk PS sample, while the segments located near the PS/PMANa interface were characterized by a restricted mobility due to the neighboring ionic groups that were thought to experience ionic association as in ionomers.

Other examples of chain dynamics measurements can be found in Hamley's book [2].

2.6 Cross-Linked Micelles

It has been shown that block copolymer micelles are dynamic structures, although they can be kinetically frozen. Unimers can thus escape from micelles and be exchanged with other micelles or be adsorbed on another interface such as the surface of a pigment particle. The unimer-micelle equilibrium is influenced by the copolymer concentration. In this respect, micelles can disintegrate into unimers if the concentration is below the CMC. Moreover, block copolymer micelles will be disassembled whenever the initial solvent is changed for a better solvent for the core-forming chains. Similarly, pH, ionic strength, or temperature changes can modify the aggregation state for the stimulus-reponsive micellar systems.

An obvious way to stabilize block copolymer micelles consists in the cross-linking of the micellar core or corona. Several strategies have been developed to reach this goal, as briefly illustrated in the following discussion.

Block copolymer micelles containing PB cores were cross-linked either by UV or fast electron irradiation [79–81]. This was accompanied by a shrinkage of the micelles.

Liu and coworkers systematically used photo-cross-linking to stabilize micelles containing a PCEMA core [82]. The micellar characteristic features (Z, etc.) were not affected by the cross-linking process, as proven by SLS, DLS, TEM, and SEC.

Wooley and coworkers have cross-linked the micellar corona and obtained the so-called shell cross-linked knedellike micelles [83, 84]. This strategy was further applied to a wide variety of block copolymer micelles. Armes and coworkers have used a similar approach for the preparation of shell cross-linked micelles with hydrophilic core and shell [85]. Many other related examples can be found in the literature.

3 Micelles from Block Copolymers in Organic Solvents

Excellent reviews on micelles formed in organic solvents have been published by Hamley [2], Chu et al. [86], and Riess [14]. From these overviews it appears that a wide range of styrene-, (meth)acrylates-, and dienes-based block copolymers were investigated and that the formation of micelles in organic solvents can generally be considered as an entropy-driven process. AB diblock and ABA triblock architectures were systematically compared. All these previous investigations have been summarized by Hamley [2]. We will therefore not perform an extensive review of all these systems, since this information has already been provided by others, but we will briefly outline some selected examples.

As far as micelles in organic media are concerned, two types of block copolymers can be considered—those with two hydrophobic blocks and those with one hydrophilic and one hydrophobic block. The latter form the so-called reverse micelles, which contain a hydrophilic core surrounded by a soluble hydrophobic corona.

3.1 Micelles from Double-Hydrophobic Copolymers in Organic Solvents

The micellization behavior of copolymers containing two hydrophobic blocks, or double-hydrophobic block copolymers, has been shown to be mainly controlled by the solvent and its interaction with the copolymer blocks. It is thus possible to tune the micellization of these copolymers by changing the organic solvent. In this respect, large differences in Z, R_h, R_c, etc. are expected whenever the interaction parameter between the polymer and the solvent is varied. This is illustrated by, e.g., the work of Pitsikalis et al. [87] for PS-PSMA diblock copolymers dissolved in either ethyl- or methylacetate. The effect of temperature has been studied by Quintana et al. [88, 89], who have clearly shown that CMC decreases with increasing temperature for PS-PEB copolymers in alkanes.

Supercritical carbon dioxide ($scCO_2$) can be considered to some extent as an organic solvent. Various chemical reactions have recently been considered in $scCO_2$, the main advantage being that the final products can be recovered by easy elimination of the "nonharmful" solvent. In this respect, "green" polymerization reactions have been considered in $scCO_2$. However, a few monomers and polymers are soluble in $scCO_2$, most of them belonging to the perfluorinated compound and silicone families. Block copolymer micelles in $scCO_2$ recently became a field of interest due to their capability to stabilize emulsionlike polymerization systems. Therefore, the micellization behavior of block copolymers containing either a perfluorinated or a polysiloxane block associated to another block compatible with the dispersed phase has been studied in $scCO_2$. Specific experimental setups have been built up to allow the study of micelles in $scCO_2$. This is illustrated by the works of McClain et al. on the micellization of PVAC-PTAN copolymers in $scCO_2$ [90]. A combination of DLS and SLS allowed the phase diagram of such copolymers in $scCO_2$ to be determined. Three regions were observed: (1) two-phase region at low CO_2 density, (2) solutions of spherical micelles at intermediate CO_2 densities, (3) solutions of unimers at high CO_2 densities. Z was found to decrease with an increasing density of $scCO_2$ in region (2) because an increase in the CO_2 density corresponds to an improvement in solvent quality for both blocks of the copolymer. SANS has been further used by the same authors to study the unimer–micelle transition in the PVAC-PTAN copolymers, which was referred to as a critical micelle density (CMD) by analogy with the CMC [91].

3.2 Micelles from Amphiphilic Copolymers in Organic Solvents

Reverse micelles from amphiphilic block copolymers have been widely investigated. In such micelles, the core is typically formed from PEO, PMAA,

PAA, or P2VPQ hydrophilic blocks while the corona consists of hydrophobic chains.

PEO-based copolymers have received much attention. In this respect, PEO-PPO and PEO-PPO-PEO Pluronic copolymers were investigated in organic solvents such as formamides, as illustrated by the works of Lindmann and coworkers and Alexandridis et al. [92–94]. However, the formation of reverse micelles in organic solvents from PEO-based block copolymers has been shown to be a complex phenomenon due to the ability of PEO to crystallize.

The crystallization of PEO has been studied in PS-PEO and PB-PEO micelles by several groups including Gallot and Gervais [95] and Gast et al. [96–98]. Wu and Chu [99] and Guo et al. [100] demonstrated that a key parameter in the crystallization of PEO was the amount of residual water. Similarly, Gast et al. [98] showed that PS-PEO copolymers form micelles in cyclopentane having Z between 17 and 100 depending on the copolymer concentration and on the water content. In the absence of water, lamellar microcrystals or shish-kebab structures are formed by chain-folding crystallization of PEO as shown by Kovacs et al. [101] and Gast and coworkers [96]. This leads to micelles with an unusual lamellar morphology, as will be discussed in Sect. 6.

PS-PEO reverse spherical micelles have been used as nanoreactors for the synthesis of metallic nanoparticles, as shown by the works of Möller and coworkers [102] and Bronstein et al. [103].

Reverse micelles from PMAA and PAA-containing copolymers have been extensively studied by Eisenberg and coworkers [104, 105]. These authors considered the micellization of the so-called "block ionomers" formed of a major PS block linked to ionized PAA and PMAA segments. Stable spherical micelles were formed by these copolymers in organic solvents such as toluene. Their characteristic size was systematically investigated by a combination of experimental techniques including TEM, SAXS, DLS, and SLS. The micelles were shown to consist of an ionic core and a PS corona. The mobility of the PS segments located near the ionic core was found to be restricted, as discussed in Sect. 2.4.

P2VP- and P4VP-containing reverse micelles have also been widely investigated. Förster et al. reported a very detailed investigation on PS-P4VP copolymers dissolved in toluene, a selective solvent for the PS block [106]. By combining TEM, DLS, and SLS, they determined how Z was scaling with the average degree of polymerization of both the P4VP and PS blocks. PS-P4VP and PS-P2VP block copolymer micelles have been widely used as templates for the synthesis of metallic nanoparticles. To achieve this goal, precursors of metals were first loaded into the micellar P2VP or P4VP core, which was achieved by simply stirring the precursor salt in the micellar solution. Transition metals were bound either directly to the P2VP or P4VP ligand in the micellar core (e.g., $Pd(AcO)_2$, $Cd(ClO_4)_2$) or indirectly as counterion. This is the case of $HAuCl_4$, which protonates the P2VP or P4VP blocks, and therefore $AuCl_4^-$ is bound as the counterion. Even if the precursor salt is soluble in the solvent, e.g., in the case of the PS-P4VP/THF system, the equilibrium distribution of metal ions is strongly shifted to the micellar core because of the much better coordination to the soft base P2VP or P4VP compared to the hard base THF [107]. In a further step, the metal precursor has to be reduced to form the metal. This step is rather easy because the redox potential of the coordinated metal is often reduced due to a partial charge transfer from the polarizable ligands to the positively charged metals. Common reducing agents are $LiAlH_4$, $NaBH_4$, H_2N-NH_2, $LiBEt_3H$, and H_2. Semiconductor nanoparticles can also be prepared by addition of H_2S to the metal precursors to form the corresponding metal sulfides. The reduction initially leads to the formation of primary metal atoms or semiconductor particles. These particles further aggregate to form larger clusters by nucleation and growth processes, as previously reviewed by Förster and Antonietti [107]. These processes can lead either to one single colloid per micellar core or to several small colloids within a micellar core.

In a more general way, the loading of metal salts into preformed block copolymer micelles has become the most used route for the incorporation of precursors into block copolymer nanostructures because it allows precursor loading with tolerable loading times, it is quite versatile, and it is applicable to a wide variety of precursor/block copolymer/solvent systems. The accordingly synthesized polymer-coated metallic or semiconducting nanoparticles exhibit increased stability, which results in, e.g., protection against oxidation as illustrated by Antonietti et al. [108].

Various metallic nanoparticles have been prepared in the case of PS-P2(4)VP reverse micelles, as illustrated by the works of Möller et al. [109–111] and Antonietti and coworkers [112]. Very recently, Bronstein et al. reported on the synthesis of bimetallic colloids formed in PS-P4VP micelles and their catalytic behavior for the selective hydrogenation of dehydrolinalool [113].

Besides metallic nanoparticle synthesis, interaction of metal salts with P2VP or P4VP blocks has a deep effect on the micellization behavior of the corresponding PS-P2VP and PS-P4VP copolymers. Indeed, the incompatibility between the core and the shell in the micelles can be enhanced by neutralizing the 2VP or 4VP units of the core-forming block by, e.g., $HAuCl_4$. To minimize the unfavorable contacts, the core chains stretch and Z is increased. This is illustrated by the work of Möller and coworkers who studied the association behavior of PS-P2VP copolymers in the presence of $HAuCl_4$ [114]. In the case of short PS coronal blocks, i.e., crew-cut micelles, spherical micelles were formed at dilute concentration. However, upon evaporation of the solvent, cylindrical micelles were observed. In a related study on PS-P2VP micelles in toluene, a morphological transition from spherical micelles to large anisotropic objects was observed by decreasing the concentration of the block copolymer below the CMC [115].

PS-P2VP copolymers with a miktoarm starlike architecture deposited onto mica or Si wafers have been investigated by Kiryi et al. [116]. These authors studied, using AFM with molecular resolution, single-molecule conformations formed in a controlled environment (Fig. 6). In toluene, unimolecular micelles were formed with the P2VP arms in the micellar core and the swollen PS arms as the micellar shell. The reverse situation was observed in acidic water. The transition between those two inverse types of micelles was strongly modified by interaction with the mica substrate. Indeed, the micelles deposited onto mica from acidic water were trapped via P2VP extended arms. Upon treatment of the trapped micelles with toluene the PS core was swollen and PS arms gradually adapted an extended conformation, whereas P2VP trapped arms retained their extended conformation due to the strong interaction with the mica substrate. The obtained structures exhibited a unique conformation that does not exist in any solvent and could not be obtained upon a simple adsorption procedure.

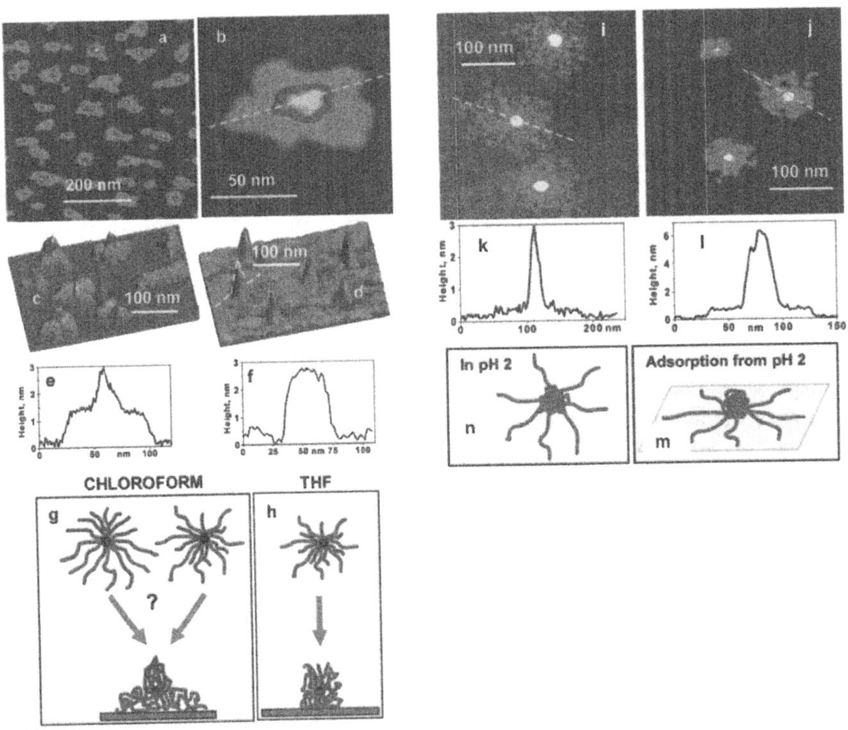

Fig. 6 AFM topographic images (**a–d**, **i**, **j**) and cross sections (**e**, **f**, **k**, **l**) of a miktoarm PS-P2VP star copolymer adsorbed on mica from chloroform (**a–c**, **e**), from THF (**d**, **f**) and from acidic water (HCl, pH = 2) in salt free (**i**, **k**) and in the presence of 1 mM Na_3PO_4 (**j**, **l**). Schematic representation of the solution conformations and conformations in adsorbed state of the PS-P2VP in chloroform (**g**), THF (**h**), in water at pH = 2 before (**n**) and after adsorption (**m**) respectively (PS arms in *red*, P2VP ones in *blue*). Reprinted with permission from [116]. Copyright (2003) American Chemical Society

4 Micelles from Block Copolymers in Aqueous Solution

4.1 Nonionic Amphiphilic Block Copolymers in Aqueous Solution

Nonionic amphiphilic block copolymers in aqueous solution are typically formed of a water-soluble hydrophilic block, e.g., PEO, PMVE, PNIPAM, linked to a hydrophobic block, e.g., PPO, PBO, PS, PMMA.

The commercially available PEO-PPO and PEO-PBO amphiphilic di- and triblock copolymers have been widely investigated in aqueous medium. Depending on temperature and concentration, different structures have been reported for the corresponding micellar solutions, including micelles of various morphologies and physical gels. Their micellization behavior has been studied quite extensively and summarized in the review articles of Chu and Zhou [19], Almgren et al. [117], Hamley [2], Booth et al. [21, 118], and Wanka et al. [119]. The morphology of spherical micelles formed by PEO-PPO copolymers has been studied using DLS, SLS, SAXS, and SANS. A rather detailed picture of the morphology of such micelles has been established, as exemplified by the work of Mortensen and Pedersen [120], who have shown that in the case of spherical micelles the PPO core is surrounded by a dense layer of PEO segments and an outer corona of flexible PEO chains.

Recent studies on PEO-PPO, PEO-PBO di- and triblock copolymers include the works of Bahadur et al. [121], who examined the role of various additives on the micellization behavior, of Guo et al. [122], who used FT-Raman spectroscopy to study the hydration and conformation as a function of temperature, of Booth and coworkers [123], who were mainly interested in PEO-PBO block copolymers with long PEO sequences, and of Hamley et al., who used in situ AFM measurements in water to characterize the morphology of PEO-PPO micelles [56, 57].

PS-PEO di- and triblock copolymers have been investigated in detail, as summarized in the review of Riess [14]. For PS-PEO diblock, Z was found to increase with the copolymer MW at constant composition and to decrease with PEO content for a given MW [124]. It was further observed that PEO-PS-PEO triblocks had lower Z than the corresponding PS-PEO diblocks [125, 126]. These micelles were, however, characterized by two typical characteristic features: (1) nonequilibrium "frozen" micelles were formed due to the high T_g PS core, and (2) these micelles were prone to secondary aggregation. Aggregation of PS-*b*-PEO micelles has been reported by, e.g., Khan et al., who observed two populations of objects by TEM [127]. Two populations, with D_h at 40 and 150 nm, respectively, were also detected by Xu et al. by light scattering in similar samples [128]. The smaller population was attributed to regular micelles, while the larger one was believed to consist of loose micellar clusters. Mortensen et al. have studied a PS-PEO copolymer using SANS, DLS, and SLS and have evidenced the formation of anisotropic clusters in water at concentrations up to 10 wt % [129]. These authors considered the clusters to be the result of the merging of the initial spherical micelles into bigger aggregates. Clustering of PS-PEO micelles in water was also investigated by Bronstein et al. using SLS and AUC [130]. These authors showed that the micellar clusters could be decomposed by the addition of toluene due to an increased mobility in mobility of the PS core-forming chains. Addition of inorganic salts interacting with the PEO blocks also resulted in the disappearance of the micellar clusters. Secondary aggregation was very recently studied in details in metallosupramolecular PS-PEO micelles, as will be discussed in Sect. 7.5 [48, 131, 132]. A cryo-TEM investigation has confirmed that the large aggregates result from the clustering of the initial spherical micelles [48]. Addition of small amounts of salt was also reducing the secondary aggregation for these copolymers [131]. Secondary aggregation was also observed for PEO homopolymers, as shown by Burchard et al. [133]. It was suggested that forces leading to clustering of PEO chains could involve hydrogen bonding, the structure of water, and the hydrophobic effect. The aggregated PEO chains were then thought to form either spherulites or non-crystalline microgel particles.

Finally, the formation of crew-cut micelles with spherical and various other morphologies was reported by Eisenberg and coworkers for PS-PEO diblocks with a major PS block [134]. More information about crew-cut micelles will be given in Sect. 6.

Nonionic hydrophilic PEO blocks have also been combined with a variety of other hydrophobic blocks, including PI [135], poly(amino acids) [136], aliphatic polyesters [137], etc. It is not possible to review all the published works on these copolymers due to space limitations. We will therefore only present a selected example.

Among PEO-containing amphiphilic copolymers, PEO-PCL copolymers have attracted much interest because they are biocompatible and partly biodegradable (for the PCL block) and they are thus ideal candidates for application in, e.g., drug delivery. A typical example on the use of PCL-PEO block copolymer micelles as delivery vehicles for drugs is found in the recent work of Luo et al. [138]. The same group recently reported on the cellular distribution of PCL-PEO micelles whose core had been labeled by a fluorescent dye. Confocal microscopy in living cells revealed the localization of the fluorescent-labeled block copolymer micelles in several cytoplasmic organelles, including mitochondria, but not in the nucleus [139]. Moreover, micelles changed the cellular distribution and increased the amount of a drug delivered to the cell [139]. This study therefore emphasized the possibility for PCL-PEO micelles to selectively deliver drugs to specific subcellular targets.

Water-soluble thermoresponsive PNIPAM and PMVE blocks have been associated to hydrophobic blocks [140, 141]. Micelles have been prepared in water in which the hydrophobic core is surrounded by PMVE or PNIPAM coronal chains. Since both PNIPAM and PMVE blocks exhibit a LCST, the coronal chains collapse upon increasing temperature and the micelles have a tendency to form large colloidal aggregates. The initial micelles are usually restored whenever the temperature is lowered below the LCST. This typical thermoresponsive behavior is illustrated in, e.g., the work of Tenhu and coworkers on micelles formed in water by PS-PNIPAM and PtBMA-PNIPAM copolymers [140].

4.2
Amphiphilic Block Copolymers with One Ionic Block in Aqueous Solution

Block copolymer micelles with a polyelectrolyte corona are a very important class of colloidal particles in aqueous medium and are often referred to as polyelectrolyte block copolymer micelles. The micellization behavior of these charged micelles has been very recently reviewed by Riess [14] and Förster et al. [15]. A brief overview of the topic will therefore be presented in what follows. Amphiphilic block copolymers consisting of one hydrophobic block linked to one ionic block will only be discussed in this section. Blocks copolymers containing one hydrophilic block and one ionic block will be discussed in Sect. 4.3.

The interest in these block copolymer micelles arises from the polyelectrolyte coronal block whose intrinsic properties are strongly influenced by many parameters including pH, salt concentration, and polar interactions. Moreover, they provide a unique model to mimic polyelectrolyte brushes at a high segment concentration, as noted by Förster [15].

Typical examples of polyelectrolyte micelles have been reported for "cationic" PS-P4VPQ and for other quaternized P2VP- or P4VP-containing block copolymers by Selb and Gallot [142, 143]. Eisenberg et al. [144–147] and Tuzar and coworkers [41, 148] have studied the PS-PAA and PS-MAA systems in their acidic or neutralized "anionic" form. Typical polyelectrolyte behavior was detected for these types of micelles.

Eisenberg and coworkers investigated highly asymmetric PS-PAA copolymers with a long PS block and a very short PAA sequence [36]. These compounds led to the so-called crew-cut micelles, as introduced in Sect. 2.3. The morphology of the crew-cut micelles could be changed from spheres to rods to vesicles and other complex micellar structures by decreasing the PAA/PS ratio in the starting copolymer [28, 36], by adding salt [149, 150], or by changing the initial nonselective solvent [38]. The underlying thermodynamic and kinetic principles explaining the formation of such morphologies were thoroughly discussed by the same authors [151]. Some of the intricate morphologies observed for crew-cut micelles are shown in Fig. 7.

As exemplified above, most of the studies on anionic polyelectrolyte block copolymer micelles have been carried out on P(M)AA-containing copolymers. The ionization degree of these anionic blocks is strongly dependent on pH: at low pH, the P(M)AA sequence is in the acidic form, while at higher pH it is partially or totally ionized. This feature could be a drawback for some applications in which a pH-independent ionization behavior is desirable. This situation could be circumvented by the use of anionic polymers derived from strong acids like PSS or PGMAS. PSS-containing block copolymer micelles have been investigated by several groups including Guenoun et al. [152, 153] and Förster and coworkers [154, 155]. PGMAS-containing micelles have been scarcely studied. We have investigated the micellization of PMMA-PGMAS and PtBMA-PGMAS copolymers in water [156]. Formation

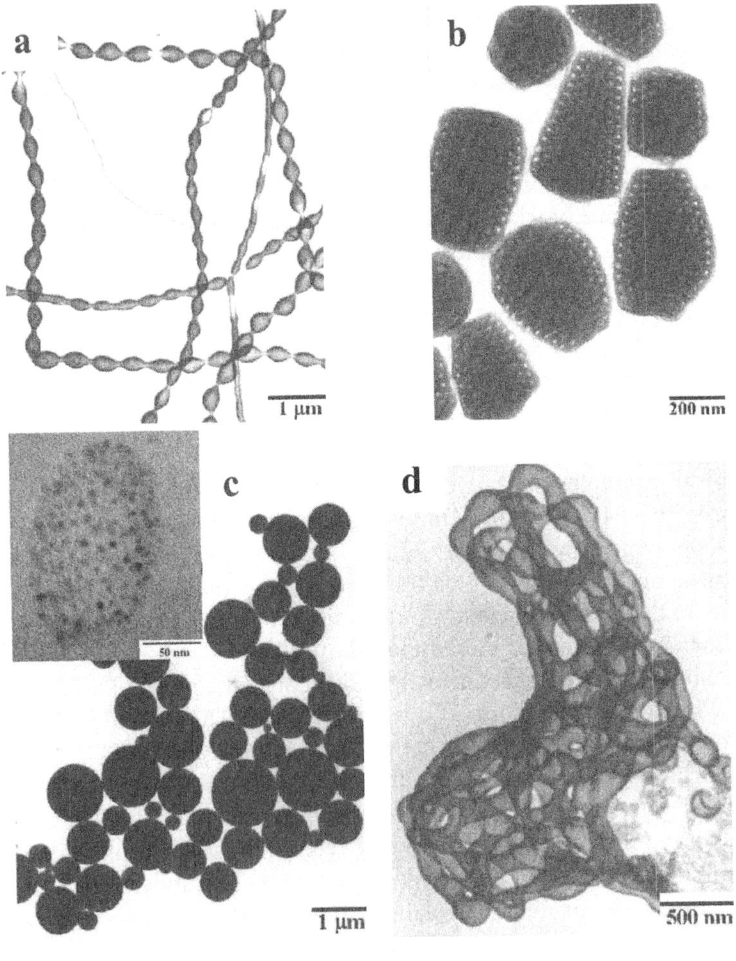

Fig. 7 Some peculiar morphologies observed for "crew-cut" micelles. Baroclinic tubes (**a**). Tube-walled vesicles (**b**). Large compound micelles; insert shows their internal structure (**c**) and interconnected tubules or "plumber nightmare" (**d**) (images downloaded from http://ottomaass.chem.mcgill.ca/groups/eisenberg/). Adapted from [35]

of superaggregates of micelles was evidenced in these samples by a combination of DLS and size-exclusion chromatography experiments on the aqueous micellar solutions. PMMA-PGMAS copolymers with a short PMMA block were also recently adsorbed on hydrophobized mica surfaces in order to create polyelectrolyte brushes showing remarkable lubrication properties upon shearing in extreme confinement [157].

Besides the pioneering works of Selb and Gallot on cationic PS-P4VPQ copolymers, more recent examples on cationic block copolymer micelles include the works of Armes et al. on various amino-containing poly(methacrylates) [158–161] with stimulus-responsive properties that were combined together or with "classical" hydrophobic blocks. Some of these copolymers could have interesting applications for pigment stabilization, as demonstrated by Creutz et al. for copolymers containing PDMAEMA and other aminated blocks [64, 65].

Dautzenberg et al. reported on an alternative method to produce cationic amphiphilic block copolymers starting from a poly(vinylbenzyl) precursor block, which was then converted into a cationic polyelectrolyte by reaction with tertiary amines [162].

Some insight about the typical polyelectrolyte behavior of cationic block copolymer micelles was recently obtained, as detailed in the review of Förster [15]. Indeed, a recent cryo-TEM study on PB-P2VP micelles in which the P2VP block was reacted with HCl revealed that the micellar PB core was surrounded by a thin dark layer containing a high density of condensed counterions, since the electronic contrast directly originated from the concentration of chlorine ions in this experiment. Furthermore thin filaments consisting of protonated P2VP chain bundles were observed that extended from the interior shell into the dilute outer part of the corona. Micelles were connected by these filaments forming a random filament network (Fig. 8).

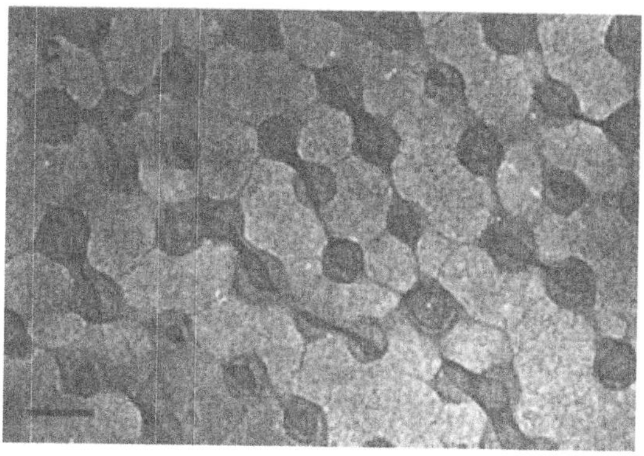

Fig. 8 Cryo-TEM image of aqueous PB-P2VPQ micelles showing filament network of polyelectrolyte chain bundles. *Scale bar* is 50 nm. Reprinted with permission from [15]. Copyright (2004) Springer

The formation of such structures was attributed to short-range attractive forces arising from transient fluctuations in the periphery of the corona, leading at times to attractive interactions between micelles. These aggregated states, i.e., strings and networks, were observed whenever the added salt concentration was increased. Further increase of salt concentration eventually led to very large networks and macrophase separation into a dilute micellar phase and a concentrated gel phase [15].

Ordered body-centered cubic structures were observed by shearing aqueous gels made from anionic PtBS-PMANa block copolymer micelles [163]. The emergence of the ordered gel state could be accounted for similar building up of a polyelectrolyte-based fibrillar network that can be oriented under shear.

4.3 Double-Hydrophilic Block Copolymers in Aqueous Solution

AB block copolymers containing two water-soluble blocks are often referred to as "double-hydrophilic" block copolymers. These copolymers have spurred much interest in recent years because they can generally be transformed into amphiphilic copolymers once an adequate stimulus is applied. In other words, one of the hydrophilic water-soluble blocks can be transformed into a hydrophobic block whenever a given property of the aqueous medium is changed. Since these copolymers may contain one or two polyelectrolytic blocks, the typical features of polyelectrolytes described in Sect. 4.2 should also be considered for some double-hydrophilic block copolymers.

Double-hydrophilic block copolymers have been successfully used as surface modifiers provided that one of the soluble blocks can specifically interact with the surface through, e.g., hydrogen-bonding interactions. Double-hydrophilic block copolymers have also been used as templates to control the crystallization of inorganic salts. Double-hydrophilic block copolymers have been recently reviewed by Cölfen [164]. In what follows, we will focus on the stimulus-responsive properties of double-hydrophilic copolymers. Stimulus-responsive block copolymers are based on polymer blocks whose water solubility can be tuned by a variation in pH, temperature, or ionic strength of the surrounding aqueous medium. Typical examples are copolymers containing weak polyacid or polybasic blocks that can be changed from hydrophilic to hydrophobic depending on their ionization degree, block copolymer micelles that contain interpolyelectrolyte complexes as the core (Sect. 7.4) and that can therefore be disassembled by screening electrostatic interactions and copolymers containing blocks with a lower or upper critical solubility temperature.

An early illustration of pH-responsive micellization can be found in P2VP-PEO copolymers that exist as unimers at pH < 5 and form micelles at higher pH values, as demonstrated by Webber et al. [165]. Recently, we investigated a P2VP-PEO copolymer with a larger P2VP block [166]. Although this copolymer precipitated out from the solution when it was totally deprotonated, micellar structures were observed in a limited pH range located near the pKa of P2VP. Moreover, a transition from spherical to rodlike micelles and finally vesicles was observed for this sample as the deprotonation degree increased, in agreement with the resulting changes in the hydrophilic/hydrophobic balance of the copolymer [166]. Giant vesicles with diameters of 5–10 μm were also recently obtained by Förster et al. from PEO-P2VP copolymers [15]. These vesicles were stable down to pH 4.5, below which the P2VP block was protonated and the vesicles dissolved.

Micellization for P2VP-PEO and P4VP-PEO copolymers has been induced at low pH by the addition of noble-metal salts as shown by Sidorov et al. [167, 168]. The driving force of such a micellization is the coordination of vinylpyridine (VP) units with metal ions, which proceeds by different basic scenarios, depending on the type of metal compound and the pH of the medium [167]. For instance, Pd^{2+} cations were directly coordinated with the nitrogen atoms of the P2VP core-forming blocks, resulting in micellization within 15 min. In the case of salts containing the noble metal in the anion (e.g., Na_2PdCl_4), the interaction of micelles with metal salts took place due to ligand exchange of chlorine for VP and the process required days to complete. Micellization was also triggered by protonation of VP units with a metal-containing acid (e.g., $HAuCl_4$). Although the salt was immediately bound, structural equilibration of the micelles demanded more time. Reduction of the metal ions embedded in the P2VP-PEO and P4VP-PEO micelles finally resulted in the formation of well-defined metal nanoparticles. In the case of $HAuCl_4$-filled P4VP-PEO micelles, the subsequent reduction with hydrazine resulted in a significant fraction of rodlike micelles [168].

Another example of pH-driven aggregation is illustrated in double-hydrophilic block copolymers based on ethylene oxide, EO, and (meth)acrylic acid, (M)AA. At low pH, hydrogen bonds form between (M)AA and EO that are further disrupted by an increase in pH due to the ionization of the (M)AA units [169]. Micellization was observed for an asymmetric PMAA-PEO diblock copolymer containing a major PEO block [170]. Intra- and intermolecular hydrogen bonding between the PMAA blocks and PEO segments was thought to result into micellar core that were stabilized by a corona formed by the excess of PEO segments. These micelles were disintegrated whenever the surrounding pH was raised above 5 as a result of the ionization of PMAA. Furthermore, these micelles showed a thermoresponsive behavior related to the limited thermal stability of the PMAA-PEO hydrogen-bonded complexes. The association behavior of linear and grafted copolymers based on P(M)AA and PEO was also recently studied by Tenhu and coworkers, but the composition of these copolymers was such that no well-defined micelles with hydrogen-bonded cores were detected at low pH [171]. Aggregates formed by double-hydrophilic copolymers based on PAA and hydroxyethylcellulose were recently reported by Dou et al. [172]. In these copolymers,

PAA chains were grafted on a hydroxyethylcellulose backbone and hydrogen-bonding interactions between the two constituent blocks were reported, resulting in the formation of hollow spheres at neutral pH.

Double-amphiphilic block copolymers have also been obtained from diblock copolymers consisting of two different charged blocks. Diblock copolymers containing positive and negatively charged blocks have been referred to as polyampholytic systems. Interpolyelectrolyte complexes (IPECs) between the oppositely charged blocks are observed in these systems, which can lead to insoluble material in aqueous solution. Asymmetric copolymers with an excess of negative or positive charges can, however, give rise to water-soluble micellar sytems, in which the micellar cores are formed by the insoluble IPECs, surrounded by a corona formed by the uncomplexed segments. The first example of polyampholytic block copolymers has been reported for P2VP-P(M)AA copolymers [173]. The authors showed that the isoelectric point of these copolymers in water was controlled by the block length ratio of the two oppositely charged blocks. PDMAEMA-PMAA copolymers were also shown to behave as polyampholytic systems [174]. Since the ionization degrees of both the PDMAEMA and PMAA blocks depend on pH, the polyampholytic character is only observed in a restricted pH region in which both the constituent blocks are partially or totally ionized. An analogous behavior was observed for the P2VP-P(M)AA copolymers. In the case of the PDMAEMA-PMAA system, vesicular morphologies were observed around pH = 9 for copolymers containing a major PMAA block. A cryo-TEM image

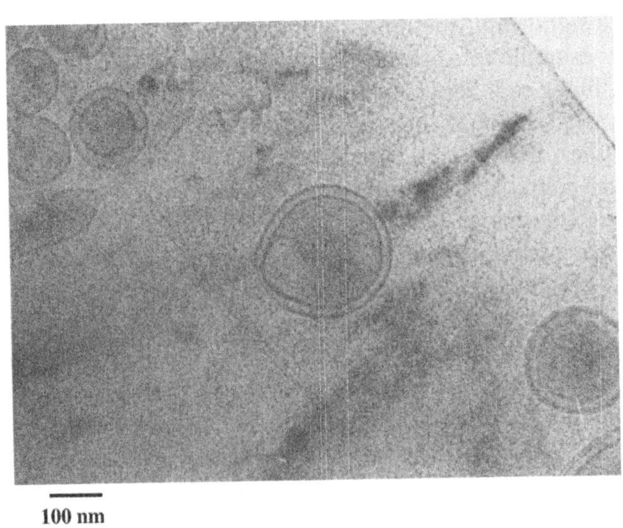

Fig. 9 Cryo-TEM picture of vesicles formed by the $PMAA_{49}$-$PDMAEMA_{11}$ ampholytic copolymer in water at pH = 9. Reprinted with permission from [174]. Copyright (2000) American Chemical Society

of these vesicles is shown in Fig. 9. These vesicles are thought to consist of an insoluble layer formed by PDMAEMA/PMAA IPECs stabilized by dangling negatively charged PMAA segments. Similar PDMAEMA-PMAA copolymers were also investigated by Cohen-Stuart et al. [175] and Armes and coworkers [176]. The adsorption behavior of a series of PDMAEMA-PMAA copolymers with various compositions and block lengths was investigated on silica by Mahltig et al. as a function of pH and ionic strength [177–180]. Nanostructured surfaces with various morphologies were obtained depending on the ionization state of silica and on the micellar structure of the adsorbed copolymer.

Polyampholytic copolymers containing a permanently charged block were reported for P2VP-PSS [181] and P4VPQ-PMAA copolymers [182, 183]. The complex solubility of P4VPQ-PMAA copolymers as a function of pH and ionic strength was investigated, and a phase diagram was presented [182].

Interpolyelectrolyte complexation between various charged (co)polymers or between charged double-hydrophilic copolymers and surfactants has been used as a tool to generate well-defined micellar structures, as will be discussed in more detail in Sect. 7.4.

Another class of charged double-hydrophilic block copolymers is found in diblock copolymers containing two charged blocks, both of them carrying the same type of charge. This situation is exemplified by PSS-PSCNa copolymers, which form micelles at low pH whenever the PSCNa block is protonated and therefore less soluble than its ionized form [184]. Double-hydrophilic block copolymers containing two basic blocks have also been studied, as reported by Armes and coworkers for PDMAEMA-PDEAEMA copolymers [185] and Gohy et al. for P2VP-PDMAEMA diblocks [186]. pH-induced micellization was observed for these two types of block copolymers. However, the relatively large difference between the pKas of the P2VP and PDMAEMA blocks allowed the formation of three different association states in aqueous solution depending on pH: at low pH, both the P2VP and PDMAEMA blocks were protonated and water-soluble; at intermediate pH, polyelectrolyte block copolymer micelles were formed with an insoluble P2VP core and a positively charged PDMAEMA corona; at high pH, micelles were formed with the same insoluble P2VP core and thermoresponsive uncharged PDMAEMA coronal chains. The adsorption behavior of these micelles on silica was also studied [187].

Selective betainization of the PDMAEMA block of PDMAEMA-PDEAEMA, PDMAEMA-PDPAEMA, and PDMAEMA-PMEMA copolymers by propanesultone resulted in thermosensitive systems, as reported by Armes and coworkers [160, 188].

Thermosensitive double-hydrophilic block copolymers micelles have been reported essentially for systems containing blocks exhibiting a LCST in water such as PMVE [189] or PNIPAM. These blocks have been generally associated to PEO blocks. At temperatures below the LCST, both blocks are soluble and no aggregation is observed. At temperatures above the LCST, the thermosensitive blocks turn hydrophobic and aggregation (micellization) takes place. An early example of temperature-induced micellization has been reported for PEOVE-PMOVE, which exhibited a multistage phase separation, as monitored by turbidimetry measurements [190]. Another typical example of such a thermoresponsive system has been reported for micelles formed by PEO-PNIPAM copolymers in water [191]. PMVE-PVA copolymers had a similar behavior, and micellization was reported to occur near the cloud point of the PMVE block at 29 °C [192]. DLS, ^{1}H NMR, and dye solubilization were used to monitor the temperature-dependent micellization of PMVE-PMTEGVE copolymers in water [193]. Unimers were detected below the LCST of the PMVE block, while well-defined micelles and a few superaggregates were present above the LCST. Moreover, the LCST values for the PMVE-based copolymers were strongly influenced by the relative block lengths of the two blocks.

Recently, more complex thermoresponsive copolymers consisting of two blocks with each presenting a LCST were investigated. In addition, these systems were sensitive to pH and ionic strength, paving the way to multiresponsive block copolymer micelles. In this respect, Armes and coworkers reported on micelles based on PPO-PDEAEMA copolymers and exploited the different pH dependencies of the LCSTs of both blocks [77]. This further allowed the switching of the inside and outside blocks by combined pH and temperature stimuli, leading to so-called schizophrenic micelles. Another interesting example was reported by Laschewsky and coworkers for block copolymers combining blocks with either a LCST or a UCST [194]. In this case, the inside and outside blocks of the micelles were changed by a simple thermal stimulus.

5 Rationalization of Micellar Structure: From Theory to Experiments

In Sects. 3 and 4, micellization in organic and aqueous media of various block copolymers containing either hydrophobic or hydrophilic nonionic polymer blocks or charged polyelectrolyte blocks has been reviewed. Spherical micelles have been formed for the vast majority of these systems. One of the major concern has been to evaluate how the characteristic dimensions of these micelles, i.e., R_c, R_m, b and Z (see Sect. 2.3), varied with the composition of the starting copolymer, i.e., N_A and N_B in the case of an AB diblock copolymer, with A being the core-forming insoluble block and B the corona-forming soluble block. This key problem has been investigated for a long time from both the theoretical and experimental points of view. Theories on block copolymer micelles have been reviewed by several authors including Tuzar [6], Hamley [2], Gast [195], and recently Linse [196] and Riess [14]. In the present review, we will briefly summarize the main results obtained from the theoretical investigations on block copolymer micelles. Theories on block copolymer micelles have also focused on the CMC, on the density profile and the thickness of the core-corona interface, and on the encapsulation/release of various species such as drugs or core-selective solvents. We will not discuss these investigations in what follows. Information about these specific issues can be found in the reviews of Hamley [2] and Riess [14].

Theories predicting how the characteristic size of block copolymer micelles depends on the composition of the copolymer are based, on one hand, on the scaling concepts derived from the Alexander–de Gennes theories, and, on the other hand, on the mean field theories developed by Noolandi and Hong [197], Leibler et al. [198], Nagarajan and Ganesh [199], and Hurter et al. [200]. In addition to theory, computer simulations have also been carried out on block copolymer micellar systems, as illustrated by the works of, e.g., Binder and Müller [201, 202] and Mattice and Haliloglu [203]. For micelles containing polyelectrolyte coronal blocks, specific theories have been developed, as illustrated by the works of Marko and Rabin [204], Dan and Tirrell [205], and Shusharina et al. [206]. They predict a strong influence of the polyelectrolyte blocks on the micellization behavior.

In scaling theories, R_c, R_m, and Z are directly correlated to N_A and N_B for the investigated micelles. Two limiting cases have to be distinguished, the starlike or hairy micelles with $N_A < N_B$ and the crew-cut micelles with $N_A > N_B$ (Sect. 2.3).

Daoud and Cotton have developed a scaling model for hairy micelles and have found that the micelle total radius, R_m, scales as $N_B^{3/5}Z^{1/5}$ [207]. Since $Z \sim N_A^{4/5}$, $R_m \sim N_A^{4/25}N_B^{3/5}$.

A similar scaling law was obtained by Halperin for a starlike micellar model [208], which demonstrates the predominant contribution of the coronal chains to the total micelle size.

For crew-cut micelles, Zhulina and Birshtein demonstrated that the dependence of the micellar parameters on N_B disappears and R_c scales as $\gamma^{1/3}N_A^{2/3}a$ and $Z \sim \gamma N_A$, where γ is the interfacial tension between the A and B blocks and a is the segment length [209].

Scaling theories are restricted to long polymer chains in good solvents and do not include finite chain effects and polymer-solvent interactions. These models should be complemented by more detailed mean-field calculations and molecular simulations.

Semianalytical mean-field theories of block copolymer micellization were formulated by Noolandi et al. [197] and by Leibler et al. [198]. In the approach of Noolandi et al., the micellar characteristics were obtained through a minimization of the Gibbs free energy for an isolated micelle. This was applied to PS-PB micelles, and the obtained theoretical values were in good agreement with the experimental ones.

A further development in mean-field theories was achieved by Nagarajan and Ganesh [199]. These authors have shown that $R_c \sim N_A^{0.73}N_B^{-0.17}$ and

$Z \sim N_A^{1.19} N_B^{-0.51}$ for PPO-PEO micelles in water. In contrast to previous results, Nagarajan and Ganesh demonstrated that the coronal B block can have a strong influence on the micellar characteristic features such as R_c, especially whenever the solvent is very good for the B block. By extending this concept, Nagarajan and Ganesh could obtain "universal" correlations for R_c, R_m, and Z as a function of N_A, N_B, the interaction parameter between the B block and the solvent χ_{BS}, the interfacial tension between the A block and the solvent γ_{AS}, and the molar volume of the solvent ν_s [199].

Monte Carlo simulations are additional methods for the study of block copolymer micelles as reviewed by Binder and Müller [202] and by Shelley and Shelley [210].

Extensive experimental characterization has been carried out by many groups in order to determine how the characteristic dimensions of block copolymer micelles depended on the composition of the starting copolymers. TEM and SAXS were two of the mainly used methods for the determination of R_c and R_m. These results have been rationalized by Förster et al., who found that, in the case of uncharged block copolymers, the grafting distance was depending on the soluble block length as $b_0 N_B^{\beta/6}$ with $b_0 \sim 1$ nm and $\beta \sim 0.8$, while Z was scaling as $Z_0 N_a^2 N_B^{-\beta}$ with $Z_0 \sim 1$ and $\beta \sim 0.8$ [106, 211]. As shown in Fig. 10, this equation successfully describes the micelles formed by uncharged diblock- [212, 213], triblock- [214], graft- [215], and heteroarm-star copolymers [216] in organic and aqueous solvents as well as for low MW cationic, anionic, and nonionic surfactants [106, 216]. Nevertheless, this

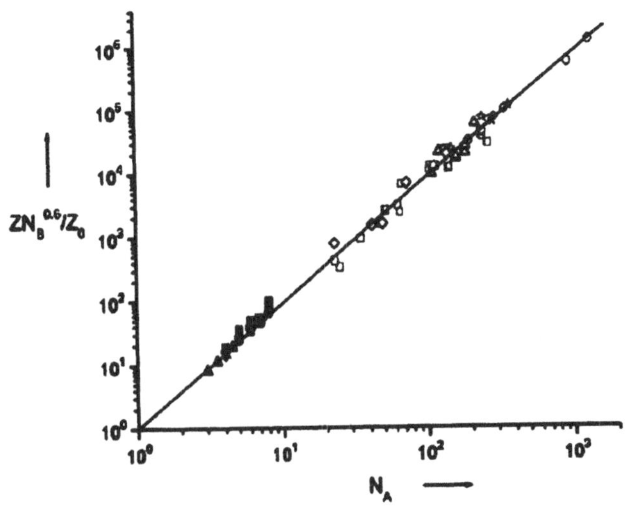

Fig. 10 Aggregation numbers Z as function of degree of polymerization of insoluble block for uncharged block copolymers. *Open symbols*: different diblock-, triblock-, graft-, and star polymers. *Filled symbols*: low-MW surfactants. Reprinted with permission from [211]. Copyright (2002) Wiley

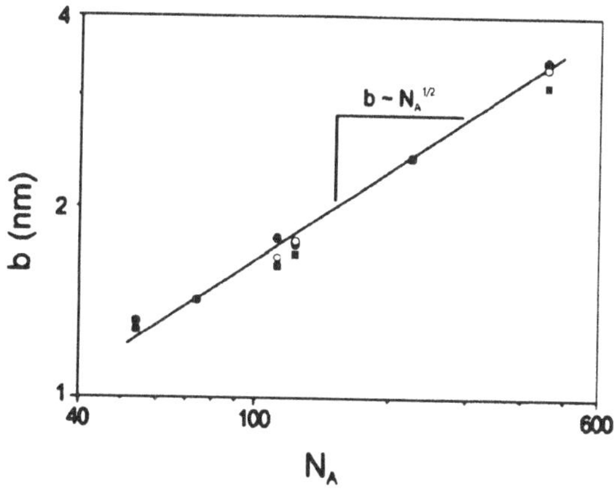

Fig. 11 Experimentally determined grafting distance b as function of degree of polymerization of insoluble block for polyelectrolyte block copolymer micelles at different salt concentrations: • salt free, ○ 0.3 mol/l, ■ 1 mol/l. Reprinted with permission from [15]. Copyright (2004) Springer

equation failed to describe micelles containing highly charged polyelectrolyte coronal blocks [15]. Indeed, the grafting distance was observed to scale as $b = b_0 N_A^{1/2}$, while the addition of salt decreases the value of R_m without affecting the 1/2-slope (Fig. 11) [15].

Salt effects in polyelectrolyte block copolymer micelles are particularly pronounced because the polyelectrolyte chains are closely assembled in the micellar shell [217]. The situation is quite reminiscent of tethered polymer brushes, to which polyelectrolyte block copolymer micelles have been compared, as summarized in the review of Förster [15]. The analogy to polyelectrolyte brushes was investigated by Guenoun in the study of the behavior of a free-standing film drawn from a PtBS-PSSNa-solution [218] and by Hariharan et al., who studied the absorbed layer thickness of PtBS-PSSNa block copolymers onto latex particles [219, 220]. When the salt concentration exceeded a certain limit, a weak decrease in the layer thickness with increasing salt concentration was observed. Similar results have been obtained by Tauer et al. on electrosterically stabilized latex particles [221].

6 Control of Micellar Morphology

In previous sections, much emphasis has been put on block copolymer micelles with a spherical morphology. It was shown in Sect. 5 that the characteristic sizes of both the spherical core and corona of block copolymer micelles can be precisely adjusted by essentially controlling the chemical nature and the degree of polymerization of the constituent blocks. For several applications of block copolymers micelles including, e.g., micellar templating in nanotechnology, it would be highly desirable to control the morphology of the micelles. For instance, one can be interested in synthesizing metallic nanorods with controlled length and diameter from cylindrical block copolymer micellar templates. It is thus not surprising that numerous research efforts have been devoted to the preparation of block copolymer micelles with various morphologies. Different strategies have been implemented as will be illustrated in what follows.

Eisenberg and coworkers have pioneered the field of micellar morphology control with the so-called "crew-cut" micelles. Generalities about the structure and preparation of crew-cut micelles have been described in Sects. 2.2,

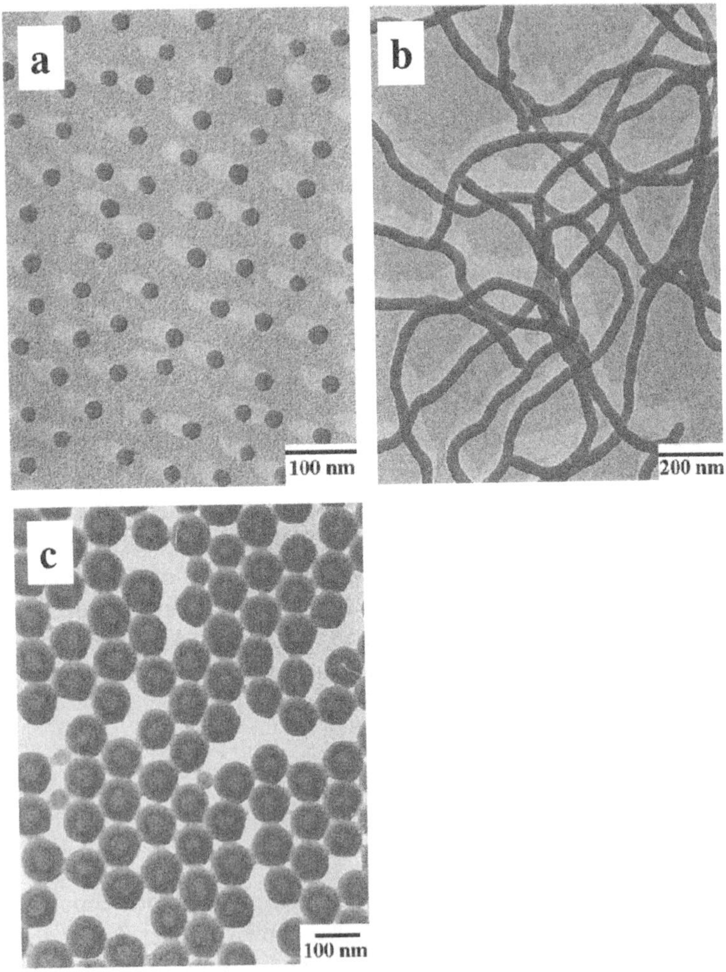

Fig. 12 Spherical (**a**), rodlike (**b**), and vesicular (**c**) morphologies for crew-cut micelles (images downloaded from http://ottomaass.chem.mcgill.ca/groups/eisenberg/). Adapted from [35]

2.3, and 4.2. These authors initially examined a series of PS-PAA copolymers with a PS block of constant length ($N = 200$), while N of the short PAA block was varied from 4 to 21. Transitions from spheres to rods to vesicles were observed as the length of the PAA block was decreased (Fig. 12).

According to these authors, the morphology of a micelle basically depends on three factors: (1) the stretching of the core-forming chains, (2) the core-corona interfacial energy, and (3) the repulsion among coronal chains. These three factors are directly related to the Gibbs free energy of the micelles, if it were the enthalpic or entropic term.

A change in one of these three parameters directly affects the free energy of the micelles. In other words, the micelles could become thermodynamically unstable and will modify their morphology in order to reach another stable state. On the basis of these arguments, Eisenberg and coworkers were able to explain morphological transitions occurring in series of PS-PAA [28, 36] and PS-PEO copolymers [134]. The effect of the starting nonselective solvent [38] and added salts [149, 150] during the micellization process on the morphologies ultimately observed were also rationalized using this simple approach [105]. The effect of the nonselective solvent on the morphology of crew-cut micelles is illustrated in Fig. 13 for PS-PAA copolymers. In addition to spheres, rods, and vesicles, other morphologies were observed by Eisenberg et al. for crew-cut micelles, including tubules, interconnected tubules called plumber nightmare, single- or multiwalled vesicles, hexagonally packed hollow hoops, large compound micelles whose cores contained inverse micelles, etc. [105]. Some of these morphologies are shown in Fig. 7.

Besides the effects of copolymer composition and preparation conditions on the observed morphologies, copolymer polydispersity can also play an important role for the formation and stabilization of some morphologies. This effect is illustrated for crew-cut vesicles formed by PS-PAA copolymers [222, 223]. In addition to the curvature energy, vesicles are characterized by a lateral strain that arises from a mismatched number of molecules in one side of the vesicular layer relative to the other. Actually, there are fewer molecules in the inner side than in the outer side of the vesicular layer. For low-MW amphiphiles, the stress resulting from this mismatch is relaxed by a "flip-flop" mechanism of the amphiphiles from one side of the vesicular layer to the other [224]. In the case of PS-PAA with a glassy wall, such a flip-flop mechanism is only possible for a fluidized PS wall (e.g., by a selective solvent). For a glassy PS wall, the stress can be minimized by segregating the long chains to the outer side of the vesicular layer while the shorter chains will preferentially localize to the inside of the vesicles. This effect has been experimentally evidenced, as reported by Eisenberg and coworkers [222, 223, 225].

One of the drawbacks of crew-cut micelles is that they systematically require the use of a nonselective solvent for their preparation and they definitely represent out-of-equilibrium micelles once they have been transferred

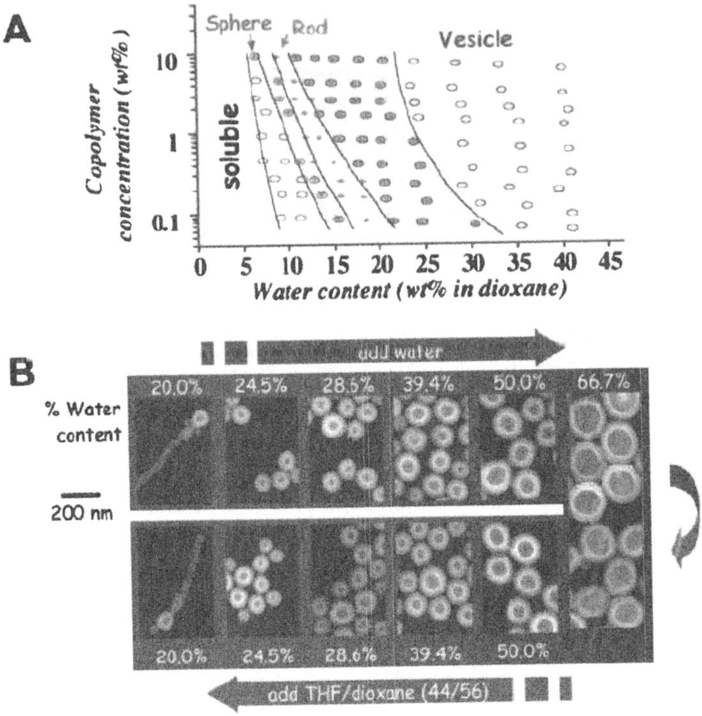

Fig. 13 Phase diagram of PS_{310}-PAA_{52} in dioxane/water mixtures (**A**). *Shaded regions* between sphere and rod phases and between rod and vesicle phases correspond to coexistence regions. Reversibility of vesicle formation and growth process for PS_{300}-PAA_{44} as function of THF/dioxane composition of nonselective solvent (**B**). Reprinted with permission from [239]. Copyright (2002) American Association for the Advancement of Science

in pure water. Micelles with rodlike or vesicular morphologies can, however, be obtained by direct dissolution of a block copolymer in a selective solvent for one of the blocks. In this respect, rodlike micelles were obtained by, e.g., Antonietti et al. [226], Liu and coworkers [227, 228], Möller et al. [114], and Bates and coworkers [229] from coil-coil block copolymers.

Discher and coworkers have investigated the formation of micelles with various morphologies by directly dissolving PEO-PB and PEO-PEB copolymers in water [230–232]. These authors especially focused on the formation of large unilamellar vesicles referred to as "polymersomes" [230]. Giant, 10-μm vesicles were prepared from these copolymers by extrusion techniques and further characterized by single-vesicle micromanipulation. This allowed the researchers to get a deep insight into the typical physical properties of these vesicles including membrane viscosity, area elasticity measurements of the interfacial tension, and electromechanical stability [231, 233]. Permeation of water through the polymersome membranes has been measured to be considerably reduced compared to phospholipid membranes [230]. Spontaneous self-organization of block copolymers into vesicles has also been reported recently for PEO-PPO-PEO copolymers with a large PPO midblock [234] and PMOXA-PDMS-PMOXA triblock copolymers terminated by cross-linkable methacrylate groups [235]. In addition, tubular micelles were very recently reported from similar PMOXA-PDMS-PMOXA triblock copolymers [236]. Vesicles were also obtained from a pentablock copolymer composed of PEO and PMPS blocks [237]. Interestingly enough, superhelical structures can also be obtained from this PEO-PMPS copolymer (Fig. 14). Furthermore, it was described that the aggregation behavior of this amphiphilic polymer can be tuned by dispersing it in water-THF mixtures. Below water concentrations of 40% (v/v) no aggregates were found. TEM demonstrated the presence of micellar rods in mixtures containing 40–80% (v/v) water. In solvent mixtures containing more than 80% water, helices with both left- and right-handed screw pitches were observed. The formation of the helical superstructure was ascribed to a cooperative process in which the handedness of the initial PMPS segment(s) determine(s) the handedness of the subsequent PMPS segments assembling in the same aggregate, such that a chiral architecture is formed by hierarchical assembly of these

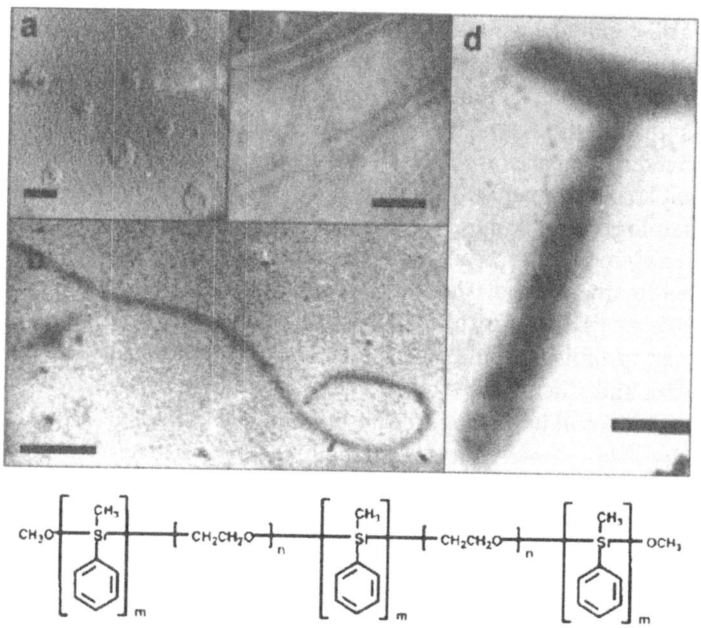

Fig. 14 TEM micrographs showing **a** vesicles, **b**, **c** micellar fibers, and **d** superhelices from a PEO-PMPS block copolymer (the chemical structure of this copolymer is represented *below* the micrographs). Reprinted with permission from [238]. Copyright (2001) American Chemical Society

dynamic helical macromolecules [238]. Superhelical micellar structures have also been observed for peptide-containing block copolymers, as discussed in Sect. 7.1.

One unifying rule accounting for block copolymer micelle morphology has been proposed by Discher and Eisenberg [239]. This rule should be considered for coil–coil block copolymer readily soluble in the selective solvent and is expressed as a function of the mass fraction of the hydrophilic block to total mass of the copolymer ($f_{hydrophilic}$).

- Spherical micelles are observed for $f_{hydrophilic} > 45\%$.
- Rodlike micelles are observed for $f_{hydrophilic} < 50\%$.
- Vesicles are observed for $f_{hydrophilic} \sim 35\%$.
- Inverted microstructures such as large compound micelles are observed for $f_{hydrophilic} < 25\%$.

Sensitivity of these rules to the chemical composition and to the MW of the copolymer chains has not yet been fully probed. According to previous experimental results, it seems that they can be applied for MWs ranging from 2700 to 20 000 g/mol.

Another possibility for triggering the formation of nonspherical micelles is to use rod–coil copolymers. Rod–coil copolymers have been recently reviewed [240, 241] and have been essentially investigated in bulk. The interest in rod–coil copolymers stems from the fact that the aggregation of the rigid segments into (liquid-)crystalline domains competes with the phase separation between the blocks during the phase-separation process. Moreover, the introduction of stiff segments results in an increase in the Flory–Huggins parameters in comparison with coil–coil copolymers [242]. As a result, phase-separated structures can be observed at lower total MW for rod–coil copolymers than for coil–coil ones.

Rod–coil copolymers can form unprecendented bulk morphologies, as illustrated by the work of Thomas, Ober, and coworkers on PHIC-PS copolymers [243, 244]. Micelle formation from rod–coil copolymers seems therefore to be a very promising route for the formation of nonspherical morphologies.

Jenekhe and Chen have studied PPQ-PS copolymers that were shown to form spherical and tubular aggregates in the 1–50 mm range with a large hollow cavity [245]. These vesiclelike structures were formed in the presence of a selective solvent of the rigid block. Large amounts of fullerenes C_{60} and C_{70} could be encapsulated in these structures [246]. Other examples of micellar aggregates from rod–coil copolymers have been reported by Stupp et al. for ABC oligomers [247], by Nolte et al. for PIC-PS diblocks [248], and by Müllen and coworkers for poly(para-phenyleneethynylene)-PDMS oligomers [249]. The first two examples will be discussed in detail in Sects. 7.1 and 7.3. In the example reported by Müllen and coworkers, micrometer-long fibrils consisting of nanophase-separated ribbons formed by rod–coil oligomers were observed.

Fig. 15 TEM micrograph of "platelet" micelles from PS_{100}-PEO_{2500} diblock copolymer dissolved in methylcyclohexane. Total size of picture: $200 \times 200\ \mu m^2$. Reprinted with permission from [14]. Copyright (2003) Elsevier

Block copolymer micelles in which the core-forming polymer blocks are able to crystallize are relatively similar to rod–coil copolymers. A significant part of these crystalline-core micelles is actually resulting from the self-assembly of rod–coil block copolymers.

PEO blocks are good candidates for the formation of micelles with crystallized cores, as demonstrated by Kovacs et al. [101] and then by Gast et al. [96]. These authors demonstrated that lamellar microcrystals or shish-kebab structures were formed by chain-folding crystallization of the PEO blocks in the absence of water. These lamellar microcrystals, called "platelets" by Gast et al., are typically formed by a central part of PEO, with a lamellar thickness depending on the crystallization conditions, having on its surface a fringe of PS blocks solubilized in organic solvent. A typical TEM micrograph of these "platelets" is shown in Fig. 15. Similar morphologies were also observed more recently by Reiter et al. [250]. Another example of crystallized cores is found in micelles formed by polyferrocenylsilane (PFS)-based copolymers, as will be discussed in Sect. 7.1 [251].

7 New Trends in Block Copolymer Micelles

Previous sections discussed the micellization behavior of AB or ABA linear block copolymers. With the recent progress achieved in the field of controlled polymerization techniques, more sophisticated block copolymer architectures are now available. Investigation of the micellization behavior of such complex architectures has been conducted in recent years. In addition, new strategies to trigger micellization have been developed. These approaches are based on the use of noncovalent interactions, e.g., electrostatic interactions, hydrogen bonding, and metal–ligand complexes, as a driving force toward the aggregation of mutually interacting polymer segments. Whenever these polymer segments are linked to other soluble noninteracting polymer blocks, well-defined micelles can result from the self-association of the interacting segments, which then become insoluble and aggregate into micellar cores stabilized by the soluble noninteracting blocks. Typical examples are based on the use of IPECs as core-forming entities. Furthermore, either the interacting segments can be of a polymeric nature or one can be a polymer block while the other is a low-MW molecule such as a surfactant.

According to these considerations, different systems have been explored up to date including:

a. AB and ABA block copolymers with nonlinear structures, such as comb-like copolymers, star-shaped copolymers, etc.
b. Functionalized block copolymers with typical structures like ABF or AFB, where F is a functional group.
c. Linear ABC triblock copolymers.
d. "Hybrid" AB block copolymers containing a synthetic polymer block linked to, e.g., an inorganic polymer, a synthetic or natural polypeptide, etc.
e. Micellization triggered by noncovalent interaction between mutually interacting polymer blocks or between a polymer block and a low-MW molecule.
f. Supramolecular block copolymers containing a metal-ligand complex as a linker between the blocks.
g. Micellization of amphiphilic graft copolymers that can have either a hydrophobic backbone and hydrophilic/ionic side chains or, inversely, a hydrophilic/ionic backbone and hydrophobic side chains.
h. Micellization of amphipolar polymer brushes, i.e., with amphiphilic block copolymer side chains, as illustrated by the works of Schmidt et al. on P2VP-PS cylindrical brushes [252] and Müller and coworkers on PS-PAA and PnBA-PAA brushes [15].

Some of these novel types of micellar systems have been reviewed by Riess [14] and by Förster et al. [15]. In this review, we will focus on points c, d, e, and f.

7.1
Micellization of "Hybrid" AB Block Copolymers

In this section, we would like to briefly describe the micellization behavior of "hybrid" AB copolymers, leading to remarkable structures with new potential applications. By "hybrid" AB block copolymers we mean copolymers in which at least one of the constituent blocks is not a "classical" polymer such as a polyolefine, a polyelectrolyte, a poly(oxyalkylene), etc., but rather a highly rigid oligomer, a synthetic or natural polypeptide, a metal-containing polymer, etc. Such systems have been recently prepared and have raised much interest because of their potential application either in nanotechnology or in bio-related systems. In what follows, we would like to emphasize two distinct types of "hybrid" systems: those based on amphiphilic copolymers formed from polyferrocene-containing blocks and those formed from polypeptide-containing copolymers.

Polyferrocene-containing block copolymers have been extensively studied by Manners and coworkers, and their micellization behavior has been examined in collaboration with Winnik's group [251, 253–260]. Wormlike micelles were prepared in hexane from a PFS-PDMS copolymer with a major PDMS block (PFS/PDMS block ratio 1 : 6), while hollow nanotubes resulted from the self-assembly of a similar PFS-PDMS copolymer but with a PFS/PDMS ratio 1 : 13. Flowerlike superstructures were also obtained from a PFS-PDMS-PFS triblock copolymer in hexane. Since the PFS block contains iron atoms, it was possible to directly visualize these micelles by TEM. It was shown that the resulting micelles consisted of a PFS core surrounded by a PDMS corona and that they were highly flexible. Crystallization of the PFS block was shown to be the driving force for the formation of rodlike structures. Whenever the micelles were prepared above the T_m of the PFS block, spherical micelles were observed. PFS blocks were recently connected to water-soluble PEO or PDMAEMA blocks and aqueous micelles were accordingly prepared. PFS-PDMS micelles consist of a potentially conducting PFS core surrounded by a PDMS insulating corona and are thus interesting candidates for the formation of semiconducting nanowires. These micelles can also be used as etching resists for semiconducting substrates, such as GaAs or Si, and can be further transformed to magnetic or semiconducting nanoscopic patterns. In this respect, cylindrical PFS-PDMS micelles have been positioned on the surface of a GaAs resist by capillary forces along grooves, which had been formed from electron-beam etching of the surface [254, 255]. Connected ceramic lines can be further obtained by subsequent treatment of the surface with hydrogen plasma. Lines less than 10 nm wide and more than 500 nm long were obtained using this approach [254].

Rod–coil copolymers are good candidates for the formation of rodlike or vesicular structures and can be built from polypeptide-block-containing copolymers, as previously discussed in Sect. 6.

Block copolymers containing an α-helical polypeptide block and a second coil-forming synthetic polymer block have been investigated for a long time, as reviewed by Lecommandoux and Klok [241]. Although most of these previous studies focused on the bulk morphology of such copolymers, micellization has been more recently reported for these systems. Indeed, Lecommandoux et al. prepared pH-sensitive vesicles from a PB-PGANa copolymer [261]. Aggregation in PEO-polypeptide block copolymers has been studied by Klok et al. [262]. Polypeptide blocks exhibit specific transitions such as PBLG, which undergoes a transition from a rodlike α-helical secondary structure to a random coil conformation under specific solvent conditions. Indeed, PBLG adopts a rodlike conformation in dioxane while it forms random coils in dioxane/trifluoroacetic acid (80/20 v/v). This transition was recently studied in bulk by SANS for PS-PBLG copolymers in which the PS block was selectively deuterated [263]. This conformational transition resulted in a topological change from a rod–coil to a coil–coil type structure. Such a conformational transition could be avantageneously used to control micellar characteristic features and to synthesize new types of stimulus-responsive materials.

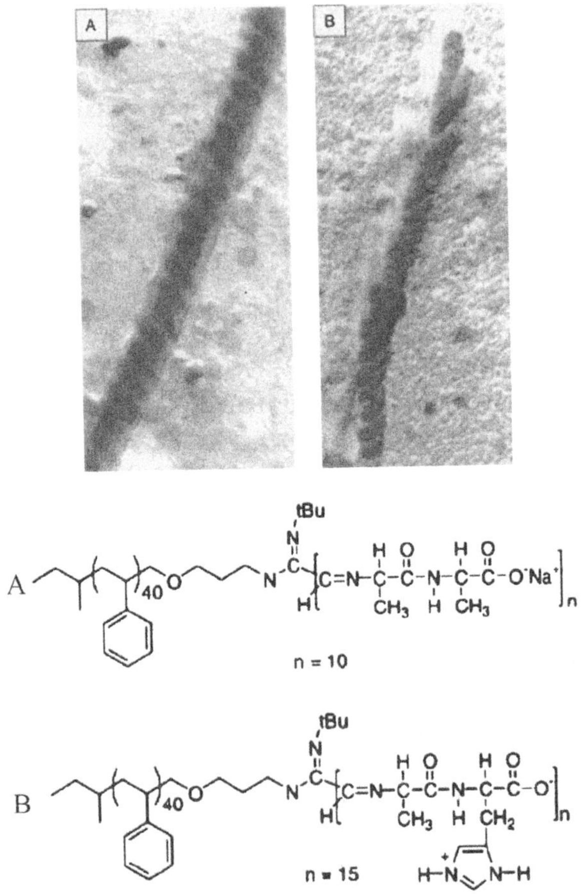

Fig. 16 TEM pictures of helical superstructures from PS-PIC copolymers in water. The chemical structure and composition of the corresponding copolymers are indicated *below* the pictures. Reprinted with permission from [238]. Copyright (2001) American Chemical Society

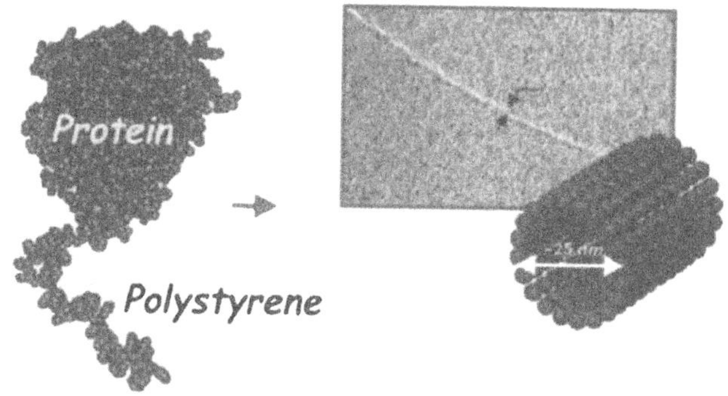

Fig. 17 Formation of rodlike micelles with diameter of 25 nm from self-assembly of a protein (lipase enzyme)-polymer (PS) hybrid in THF. Reprinted with permission from [264]. Copyright (2003) American Chemical Society

Nolte and coworkers reported on the formation of micelles with a helical superstructure from AB amphiphilic diblock copolymers prepared from an amine-end-capped polystyrene that was used as the initiator for the polymerization of various dipeptide-derived isocyanides (Fig. 16).

These authors found that the helical screw pitch of the poly(isocyanide), PIC, depended on the investigated sample. Rods, vesicles, and bilayer fragments were observed, depending on the investigated copolymers [248]. Helical superstructures were also observed, as illustrated in Fig. 16. Interestingly enough, the handedness of the superhelices had been reversed with respect to that of the polyisocyanide helices in the constituent block copolymers [248]. Although the precise mechanism of formation of these superstructures is still unclear, a hierarchical organization is evident since transfer of chirality takes place from the amino acid in the monomer to the secondary helical structure of the polyisocyanide blocks and, ultimately upon assembly, to the helical superstructure.

The same research group recently carried out the synthesis of other "hybrid" copolymers in which hydrophobic PS chains had been covalently linked to natural polypeptides such as a lipase enzyme [264]. The resulting biohybrid has been referred to as a giant amphiphile and forms catalytic micellar rods in water, as shown in Fig. 17.

7.2 Centrosymmetric Micelles from ABC Triblock Copolymers

The vast majority of block copolymer micelles has been constructed from AB diblock copolymers. However, ABC triblock copolymers have attracted a great deal of interest due to the huge number of different morphologies that have been observed so far in bulk and because the introduction of a third block may introduce interesting new functionalities. Although many investigations have been performed on bulk ABC triblock copolymers [265], these copolymers have been scarcely used for the preparation of block copolymer micelles.

In what follows, we will discriminate between two types of ABC triblock micelles: (1) those in which two blocks are insoluble in the considered solvent and then contain a compartmentalized core and a homogenous corona; (2) those in which only one block is insoluble and whose corona is heterogeneous due to the presence of two types of coronal blocks.

Micelles of type (1) were the first investigated examples of ABC triblock copolymer micelles. These micelles are generally characterized by the so-called "onion," "three-layer," or "core-shell-corona" structures, i.e., the first insoluble A block forms the micellar core, the second insoluble B block is wrapped around the core, and the third soluble C block extends in the solution to form the micellar corona (Fig. 18). To the best of our knowledge, there are no known examples of ABC block copolymer micelles with A and C insoluble blocks and a B soluble block.

Core-shell-corona micelles were formed by PEHA-PMMA-PAA triblock copolymers in water, as demonstrated by Kriz et al. [266]. Ishizone et al. [267] synthesized ABC triblock copolymers containing 2-(perfluorobutyl)ethyl methacrylate, tBMA, and 2-(trimethylsilyloxy) ethyl methacrylate with various block sequences. These copolymers were converted into amphiphilic sys-

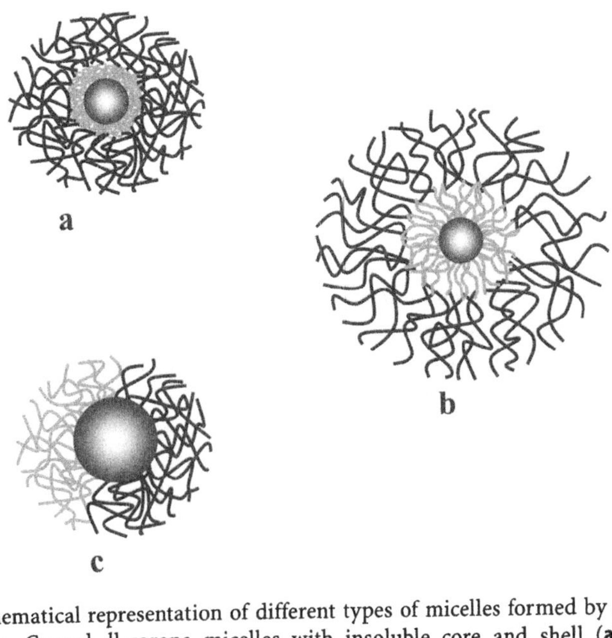

Fig. 18 Schematical representation of different types of micelles formed by ABC triblock copolymers. Core-shell-corona micelles with insoluble core and shell (**a**), core-shell-corona micelles with radially compartmentalized corona (**b**), and Janus micelles with laterally compartmentalized corona (**c**)

tems after removal of the trimethylsilyl protecting group and the formation of triblock copolymer micelles was evidenced by ^{1}H NMR measurements.

Onionlike micelles were prepared by Prochazka et al. [268] by mixing PtBA-P2VP micelles to P2VP-PEO chains in acidic water. At pH below 4.8, the PtBA-P2VP micelles consisted of a PtBA core surrounded by a protonated P2VP corona, while PEO-P2VP chains behaved as double-hydrophilic copolymers and thus existed as unimers. When the pH of the solution was raised above 4.8, insolubilization of the P2VP chains occurred. This process led to a coprecipitation of the P2VP chains from both the PtBA-P2VP micelles and the P2VP-PEO chains, which resulted in composite micelles with a PtBA core, a collapsed P2VP shell, and a PEO corona.

Similar onion type micelles were obtained by the combination of PS-P2VP heteroarm-star copolymers with a P2VP-PEO diblock copolymer, as reported recently by Tsitsilianis et al. [269].

Crew-cut micelles were prepared from PS-PMMA-PAA triblock copolymers with a large PS block [270]. The morphology of these micelles was found to be dependent on the starting nonselective solvent (dioxane, THF, or DMF), as discussed previously in Sect. 6.

Core-shell-corona micelles with a stimulus-sensitive shell were also recently obtained. We reported on micelles formed by PS-P2VP-PEO triblock copolymers in water [46, 47]. These micelles consisted of a PS core, a pH-responsive P2VP shell, and a PEO corona. At pH values above 5, the P2VP blocks were hydrophobic and collapsed on the PS core, while at pH values below 5, they were protonated and adopted a stretched conformation because of the mutual electrostatic repulsions. These two micellar states have been characterized by DLS and visualized by TEM and AFM (Fig. 3) for spherical PS-P2VP-PEO micelles. Moreover, the transition between the two types of micelles was entirely reversible and could be repeated many times. Interestingly enough, cylindrical micelles could be prepared from the same PS-P2VP-PEO copolymers using a mixture of toluene and DMF as the initial solvent before the addition of water [271, 272]. Core-shell-corona micelles were also formed, but the P2VP chains in the tubular shell were found to be in a stretched conformation whatever the pH. Similar PS-P2VP-PEO micelles were also recently studied by Prochazka et al., who evidenced the formation of clusters of micelles by FCS [273]. The pH-responsive P2VP shell was also used as template for the production of gold nanoparticles [47] and could be complexed with low-MW anionic surfactants [274]. PB-P2VP-PEO copolymers have also been investigated by Riess et al. and used as stabilizers for pigment particles in aqueous media [14].

Temperature- and pH-sensitive core-shell-corona micelles were also recently reported by Armes et al. Moreover, the shell of these CSC micelles could be selectively cross-linked [275].

pH-responsive micelles from PS-P2(4)VP-PMAA triblock copolymers were prepared by Giebeler et al. [276]. These authors investigated the polyelectrolyte complex formation in these micelles by potentiometric, conductometric, and turbidimetric titrations of acidic THF/water solutions. The formation of an IPEC at the isoelectric point was evidenced in which most likely the hydrophobic PS cores are embedded in a mixed corona of the two polyelectrolyte blocks.

Stimulus-responsive ABC triblock micelles were also investigated by Patrickios et al. for copolymers containing insoluble PEVE blocks, thermoresponsive PMVE midblocks, and water-soluble PMTEGVE outer blocks with varying block sequences [277]. While in aqueous solutions only unimers were found, and the addition of salt led to aggregates.

For some applications, it is desirable to lock the micellar structure by cross-linking one of the micellar compartments, as discussed previously in Sect. 2.6. Cross-linked core-shell-corona micelles have been prepared and investigated by several groups as illustrated by the work of Wooley and Ma [278], who reported the cross-linking of PS-PMA-PAA micelles in aqueous solution by amidation of the PAA shell. Very recently, Wooley et al. prepared toroidal block copolymer micelles from similar PS-PMA-PAA copolymers dissolved in a mixture of water, THF, and 2,2-(ethylenedioxy)diethylamine [279]. Under optimized conditions, the toroidal phase was the predominant structure of the amphiphilic triblock copolymer (Fig. 19). The collapse of the negatively charged cylindrical micelles into toroids was found to be driven by the divalent 2,2-(ethylenedioxy)diethylamine cation.

Liu and coworkers have investigated PI-PCEMA-PtBA triblock copolymers in which the central PCEMA block was cross-linked by UV irradi-

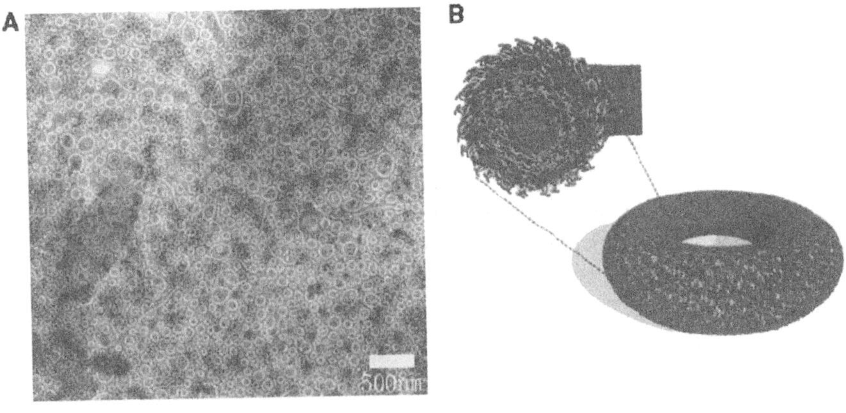

Fig. 19 TEM image of toroidal micelles from a PAA-PMA-PS triblock copolymer (**A**). This sample was cast from a solution with 0.1 wt % PAA_{99}-PMA_{73}-PS_{66} triblock copolymer, a THF : water volume ratio of 1 : 2, and an amine : acid molar ratio of 0.5 : 1 by addition of 2,2-(ethylenedioxy)diethylamine. The cast film was negatively stained with uranyl acetate. A schematical representation of theses micelles is also shown (**B**). Reprinted with permission from [279]. Copyright (2004) American Association for the Advancement of Science

ation [280, 281]. In a further step, hollow nanospheres and nanotubes were obtained by ozonolysis of the PI micellar core. Armes et al. cross-linked by a difunctional alkyl iodide the central shell of micelles formed by PEO-PDMAEMA-PDEAEMA and PEO-PDMAEMA-PBAEMA ABC triblock copolymers [282].

The micellization of ABC triblock copolymers with a starlike architecture has been reported by Dumas et al., who have studied the micellization of a PS-PMMA-PEO star copolymer in water [283]. Very recently, multicompartment micelles in dilute aqueous solution were reported for miktoarm ABC triblock copolymers containing PEO, hydrogenated PB, and perfluorinated polyether blocks [284]. Because of the miktoarm architecture of the copolymers, the two incompatible hydrophobic blocks were forced to make contact with the PEO coronal chains. The resulting micellar structures were shown to depend on the relative lengths of the blocks and could be tuned from discrete multicompartment micelles to extended wormlike structures with segmented cores. The wormlike structures were shown to result from the uniaxial clustering of discrete micelles. In this peculiar morphology, the different cores are able to share their PEO coronas, thus protecting them from the highly unfavorable exposure to water. A cryo-TEM image of such wormlike structures is shown in Fig. 20.

As introduced previously, type 2 ABC triblock copolymer micelles are formed by triblock copolymers containing an insoluble A block while the B and C blocks are soluble in the considered solvent. The insoluble blocks can be located either between the two soluble blocks (BAC structure) or at one end of the triblock (ABC or ACB structures). Micelles of the latter type were discussed above for, e.g., PS-P2VP-PEO pH-responsive micelles and are indeed considered as "core-shell-corona," "onion," or "three-layer" structures since the heterogeneity in the micellar corona is observed in the radial direction (Fig. 18). Micelles formed by BAC triblock copolymers are different from the previous case because they can give rise in principle to a heterogenous corona in the lateral dimension (Fig. 18). This could induce the formation of noncentrosymmetric micelles as discussed in Sect. 7.3.

Early examples of micelles formed by BAC triblock copolymers have been reported by Patrickios and coworkers, who studied the formation of micelles from polyampholytic PDMAEMA-PMMA-PMAA and PDMAEMA-poly(2-phenylethylmethacrylate)-PMAA copolymers as a function of pH [285–287].

Micelles from PS-P2VP-PMMA triblock copolymers dissolved in toluene were reported by Tsitsilianis and Sfika [288]. Since the organic solvent was selective for both the PS and PMMA blocks, these authors observed the formation of spherical micelles with a dense P2VP core, surrounded by PS and PMMA chains in the corona. It was shown that Z and the micellar size were strongly influenced by the length of the P2VP middle block.

Polyampholytic micelles from PDMAI-PS-PMAA triblock copolymers were reported by Bieringer et al. [289]. Investigations on the pH-dependent

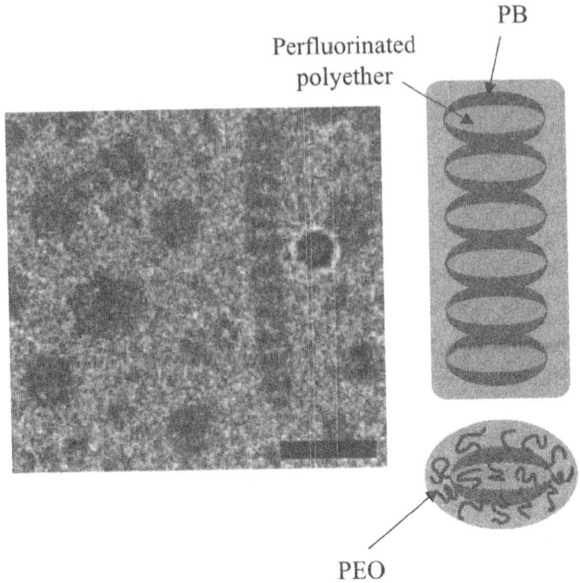

Fig. 20 Cryo-TEM image of isolated micelles and segmented rods made from miktoarm ABC block copolymers containing PEO, PB and perfluorinated polyether blocks (scale bar is 50 nm). A schematic representation of the stacking of block copolymer chains in segmented worms is also shown. Reprinted with permission from [284]. Copyright (2004) American Association for the Advancement of Science

solution properties of these micelles had to be performed in THF/water mixtures due to solubility problems in pure water. Vesicles were reported for this system. Under acidic conditions both PDMAI and PMAA blocks are soluble and the insoluble core of the vesicles is formed by the PS block, while in basic conditions both PS and PDMAI blocks are insoluble and the deprotonated PMAA block keeps the vesicular structure in solution.

7.3
Noncentrosymmetric Micelles from ABC Triblock Copolymers

ABC triblock copolymers have recently proven to be useful in constructing the so-called "three-layer," "onion," or "core-shell-corona" micelles, as described in Sect. 7.2. These micelles are characterized by a centrosymmetric structure and a micellar core with two different concentric compartments. Noncentrosymmetric structures from ABC triblock copolymers blended with AC diblocks have, however, been reported in bulk by Goldacker et al. [290].

It was recently demonstrated that noncentrosymmetric micelles can also be obtained from ABC triblock copolymers. This is illustrated by the pioneering work of Stupp and coworkers, who synthesized a "miniaturized" ABC triblock containing PS and PI oligomeric block linked to a third rodlike block formed by biphenyl esters and completed by a phenolic residue [247]. Interestingly enough, these copolymers self-assemble into nanostructures highly regular in size and shape that can be regarded as micelles whenever they are dispersed in a selective solvent for the PS and/or PI block. These nanostructures are characterized by a mushroom shape, which is thought to result from the frustrated crystallization of the rod–coil blocks. Repulsive forces among the PI and PS segments with large molecular cross section are possibly at the origin of the frustrated crystallization and, in turn, of the finite nature of these mushrooms. The mushroom micelles are thus characterized by a noncentrosymmetric structure with a hydrophilic mushroom foot due to the phenolic residues and a hydrophobic mushroom cap (PS chains). These micelles were found to further self-assemble into 100 or more macroscopic layers stacked in polar arrangement, exhibiting typical properties of noncentrosymmetric materials such as a significant second-harmonic generation [247, 291, 292].

A very recent example of noncentrosymmetric micelles has been reported for the so-called "Janus" micelles by Erhardt et al. [293]. The Janus micelles consist of a cross-linked core surrounded by two different coronal hemispheres. These micelles were obtained from dissolution of the bulk morphology of a PS-PB-PMMA triblock copolymer in which the central PB minor block forms spheres at the interface between PS and PMMA lamellae. After stabilization of the bulk structure by cross-linking of the PB spheres, Janus micelles with a PB core surrounded by PS and PMMA half coronas were obtained by dissolution of the bulk material in THF, a good solvent for both the PS and PMMA chains (Fig. 21) [293]. These Janus micelles showed a strong tendency to form larger aggregates with a narrow size polydispersity and that can be regarded as supermicelles. The supermicelles coexist with single Janus micelles, as evidenced by various morphological observations realized both in THF solution and on silicon and water surfaces [293, 294]. The hydrophobic/hydrophilic contrast in the Janus micelles can be strongly enhanced by the hydrolysis of the PMMA chains resulting in PMAA hydrophilic blocks (Fig. 21) [55]. The resulting micelles can then be gradually transferred in pure water starting from a 1,4-dioxane solution. However, the hydrophobic PS hemicorona collapses on the PB core and the whole insoluble part is then fully or partially surrounded by the PMAA chains. Aggregation into supermicelles is of course extremely favorable in this case, as proven by a variety of techniques [55]. The formation of aggregates of supermicelles was also evidenced from these investigations. A typical SEM picture showing supermicelles and their aggregates is shown in Fig. 21.

The influence of pH on PS-PB-PMAA Janus micelles was also investigated. At high pH, R_h of the supermicelles is larger than for acidic conditions, as a result of the ionization of the PMAA chains leading to their stretching [55]. Janus micelles have also been preliminarily investigated by Saito et al. [295].

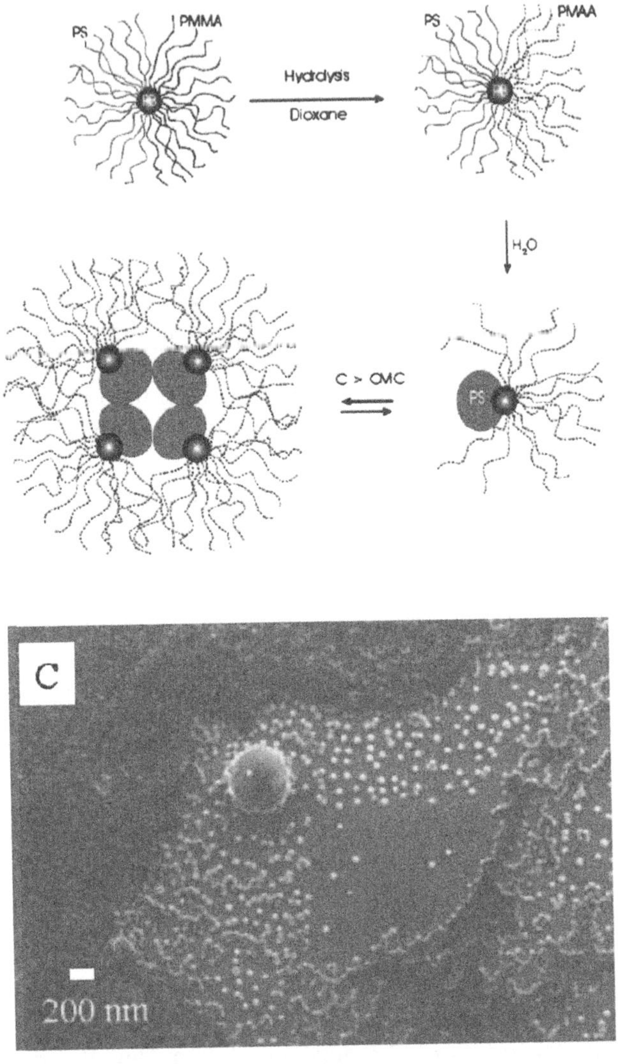

Fig. 21 Formation of Janus micelles from PS-PB-PMMA copolymers. The copolymers are transformed into PS-PB-PMAA copolymers after hydrolysis of PMMA block. PS coronal blocks collapse in water and supermicelles are formed. A typical SEM picture of supermicelles is shown. Reprinted with permission from [55]. Copyright (2003) American Chemical Society

7.4
Micelles with Noncovalent Complexes in the Core

Up to now, we have considered the micellization behavior of isolated block copolymer chains. It has been, however, demonstrated that micellization can be induced by complexation of one of the constituent blocks of the copolymer with other macromolecules or low-MW molecules. This strategy was briefly introduced in Sect. 4.3, where stimulus-responsive micellization driven by intramolecular noncovalent interactions between the constituent blocks of double-hydrophilic copolymers was discussed. This was illustrated by polyampholytic diblock copolymers containing at the same time positively and negatively charged blocks, leading to the formation of a water-insoluble IPEC. In some cases, the IPEC could self-assemble into micellar cores stabilized in aqueous solution by uncomplexed charged segments. Since the ionization degrees of both the cationic and anionic blocks were dependent on pH, stimulus-responsive systems could be built from these copolymers [174, 175].

Similar IPEC-based aqueous micelles can be obtained from the mixing of a double-hydrophilic copolymer AB with a homopolymer C or a diblock copolymer AC or a diblock copolymer DC, provided that IPECs can form between the B and C blocks. This kind of electrostatic complex has also been referred to as "block ionomer complexes," or BIC.

The AB + C case is illustrated by the works of Kabanov et al. [296], who investigated mixtures of PEO-PMANa with P4VPQ, of Gohy et al. on the pH-dependent complexation of P2VP-PEO with PSS [297, 298], and of Kataoka and coworkers on PEO-PAspA/chicken egg white lysozyme mixtures [299, 300]. In all these studies, the resulting micelles contained an IPEC core and a PEO corona.

IPEC-based micelles in aqueous solutions have also been reported from AB + AC mixtures, as illustrated by the works of Kataoka and Harada [301, 302], who have shown that water-soluble IPECs were formed by the combination of PEO-PLL and PEO-PAspA. These authors have further demonstrated that IPEC micelles prepared under charge-neutralized conditions have an extremely narrow size distribution if matched pairs of copolymers with the same block lengths of polyanions and polycations are combined [302]. Gohy et al. reported on mixtures of P2VP-PEO and PMAA-PEO copolymers [170]. IPEC formation was observed in a limited range of pH in which both the PMAA and P2VP blocks were ionized. The IPEC cores formed through electrostatic interaction between the ionized PMAA and P2VP blocks were surrounded by a PEO corona. When the pH of the aqueous medium was decreased, the PMAA and P2VP blocks were essentially observed in their protonized form. The PMAA/P2VP interaction was therefore suppressed at low pH at the expense of hydrogen bonding in the PMAA/PEO pair, as previously discussed in Sect. 4.3. As a result, micelles were formed by the PMAA-PEO diblock while the protonated P2VP-PEO chains existed as unimers. In addition, a thermoresponsive behavior was observed for the hydrogen-bonded PMAA-PEO micelles [170]. Above 45 °C, the hydrogen-bonded PMAA/PEO complexes were disrupted while the cloud point of the PMAA block was reached, resulting in stable colloidal particles. Moreover, the temperature effect was found to be entirely reversible. At higher pH, the P2VP blocks were deprotonated and hydrophobic while the PMAA blocks were essentially ionized in the form of PMANa. As a result, PMANa-PEO chains were observed as unimers while P2VP-PEO chains aggregated into micelles. In conclusion, the PMAA-PEO/P2VP-PEO mixture resulted in a multiresponsive (pH, temperature, and ionic strength for IPEC core) micellar system based on various noncovalent interactions.

IPEC-containing micelles based on double-hydrophilic cationic block copolymers, such as PLL [303–306], PEI [307], or PDMAEMA [308] in combination with PEO, have been examined for the complexation of oligonucleotides in order to provide new pharmaceutical forms for gene therapy [309].

Hydrogen-bonded complexes between mutually interacting polymer blocks can be used instead of IPEC for the formation of micelles with a complexed core. In this case, the complexation is generally realized in aprotic organic solvents in order to avoid competitive hydrogen bonding with the solvent. This is illustrated by the work of Liu and coworkers on micelles formed by mixing a PS-PMMA copolymer and a poly(styrene-*co*-[*p*-(2,2,2-trifluoro-1-hydroxy-1-trifluoromethyl)ethyl-methylstyrene]), PSOH, random copolymer in toluene [310]. Stable micelles with hydrogen-bonded cores made of insoluble PMMA/PSOH complexes and soluble PS shells were observed as long as the *p*-(2,2,2-trifluoro-1-hydroxy-1-trifluoromethyl)ethyl-methylstyrene content of PSOH was higher than 8 mol %.

IPEC or hydrogen-bonded complexation between the B and C blocks of AB and CD copolymers is a very promising route for the formation of complex micellar structures with compartmentalized corona. Noncentrosymmetric "Janus"-like micelles could in principle be obtained using this strategy. Recently, Schlaad and coworkers investigated an AB + CD system formed by PS-PMACs and PB-P4VPQ copolymers in THF [311]. However, this resulted in the formation of centrosymmetric vesicles with a vesicular layer formed by a PMACs/P4VPQ IPEC while the incompatibility between the two types of coronal chains resulted in an outer PS corona and an inner PB one (Fig. 22). It should also be noted that vesicles were prepared from a mixture of PS-PAA and PS-P4VP diblock copolymers [225]. In this case, micelle formation was not driven by IPEC formation between the PAA and P4VP blocks because micellization was carried out at low pH where the PAA blocks are protonized and the P4VP chains are positively charged. The vesicular layer was thus formed by PS chains coming from both copolymers while the PAA blocks were segregated into the inside of the vesicles and the outside corona of the vesicles consisted of P4VP chains. This segregation occurred because the PAA chains were much shorter than the P4VP chains and stabilized the curvature (Sect. 6). Electrostatic repulsion between charged P4VP blocks were also minimized.

In a very recent study, Gohy et al. reported on hydrogen-bonded complexed cores obtained in DMF through the interaction of a PS-P2VP-

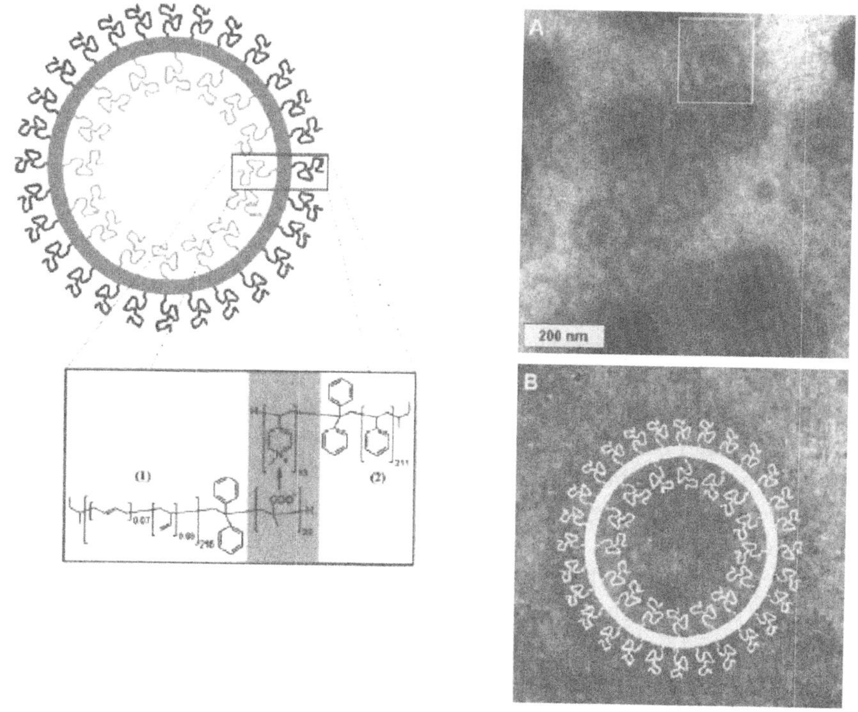

Fig. 22 Formation of vesicles with distinct inner (PB) and outer (PS) coronal blocks from a mixture of PS-P4VPQ and PB-PMACs copolymers. Vesicles are stained with CsI in the TEM pictures. TEM picture **B** is a magnification of the framed area in **A**. Reprinted with permission from [311]. Copyright (2003) American Chemical Society

PEO triblock copolymer with a tapered triblock copolymer consisting of a central PAA block and two outer comblike blocks containing short PEO branches [312]. Hydrogen bonding was observed between the P2VP and PAA blocks resulting in micellar cores while three types of coronal chains coexisted: linear PS and PEO chains and comblike PEO. Segregation was observed between the comblike PEO and the other coronal chains. Whenever the PS-P2VP-PEO triblock was complexed with a PAA homopolymer, no segregation in the corona was observed.

IPEC or hydrogen-bonded complexes may form not only between mutually interacting polymer blocks but also between a polymer block and low-MW molecules. Complexes between surfactants and block copolymers have been investigated for the formation of micelles. As illustrated by the work of Ikkala and coworkers [313], one of the major interests of these systems is that they combine two different-length scales of supramolecular organizations, i.e., the nanometer-scale organization of the (liquid) crystalline surfactant molecules and the ten-nanometer scale relative to block copolymers. This gives rise to the so-called hierarchical systems. The field of (block) (co)polymer/surfactant complexes has become an area of intense research, and it will not be possible to review it here.

Electrostatic complexes between polyelectrolytes and mutually interacting surfactants were recently reviewed by Chu and Zhou [314]. Selected examples related to block copolymer micelles will be briefly outlined in what follows. Water-soluble surfactant-block copolymer complexes were reported by Eisenberg et al. for PEO-PMANa/cationic surfactant mixtures [315, 316]. This resulted in the formation of vesicles in water with PMANa/cationic-surfactant-insoluble vesicular layers surrounded by PEO coronal chains. The characteristic features of the complexes were found to be strongly dependent on the PMANa-cationic surfactant stoichiometry, the more dense and defined structures being observed whenever all the MANa units interacted with one surfactant molecule. In a recent study, Kabanov and coworkers used the same PEO-PMANa copolymers with a reactive single-tail surfactant, isothiuroniumethyl-hexadecyldimethylammonium bromide (C16SU), which dimerizes upon cleavage of isothiuronium group at elevated temperatures or basic pH. This allowed for the reactive stabilization of vesicles from C16SU cations assembled on the anionic PEO-*b*-PMANa template [317]. We recently obtained larger vesicles while complexing a PEO-P2VPQ copolymer with negatively charged fluorinated surfactants [318].

Block copolymer/low-MW-molecule complexes were also examined in organic solvents, as recently exemplified by the works of Jiang and coworkers [319, 320]. These authors investigated mixtures of PS-P4VP copolymers with various low-MW molecules including perfluorooctanoic acid and formic acid. Such molecules are expected to form hydrogen-bonded complexes with 4VP units in organic solvents such as chloroform. This further resulted in the formation of vesicles.

Finally, micellization can also be triggered by other types of noncovalent interactions besides electrostatic complexation or hydrogen bonding. In this respect, metal–ligand coordination has proven to be an effective noncovalent interaction for promoting micellization, as illustrated by the works of Sidorov et al. [167, 168] on P4VP-PEO and P2VP-PEO copolymers, in which micellization at low pH could be promoted by metal–ligand complexation. These systems were described in Sect. 4.3.

7.5 Metallosupramolecular Micelles

Terpyridine ligands are known to form *mono-* and *bis-*complexes with a wide variety of transition metal ions [321]. The stability constants and the kinetics of formation of these different complexes strongly depend on the nature of the used metal ions [322]. In this respect, Ru^{III} is known to form a very stable *mono*-complex with one terpyridine ligand, while Ru^{II} only forms a stable *bis*-complex with two terpyridine ligands [323].

Terpyridine moieties have been introduced as a terminal unit of macromolecules. In a subsequent procedure the two-step self-assembly process based on Ru^{III}/Ru^{II} chemistry was used for polymers end-capped with the 2,2′:6′,2′-terpyridine ligand. More precisely, the terpyridine-functionalized polymers were complexed with $RuCl_3$ to selectively form a *mono*-complex. In a further step, this *mono*-complex was reacted under reducing conditions with other uncomplexed 2,2′:6′,2′-terpyridine-terminated polymer blocks in order to form an asymmetrical AB ruthenium(II) *bis*-complex.

A complete description of the synthetic methodology and the characterization of the obtained metallosupramolecular block copolymers was reported in a recent paper [324]. These compounds have been referred to as "metallosupramolecular block copolymers" and designated by the acronym A_x-[Ru]-B_y, where A and B are the two different polymer blocks, -[Ru]- denotes the *bis*-2,2′:6′,2′terpyridine-ruthenium(II) linkage between the A and B blocks, and x and y represent the average degree of polymerization of the A and B blocks, respectively. The chemical structure of a PEB-[Ru]-PEO metallosupramolecular copolymer is depicted in Fig. 23.

Compared to classical covalent block copolymers, metallosupramolecular block copolymers offer several advantages. A wide variety of polymeric blocks can be combined, regardless of the chemical structure and reactivity ratios of the constituent comonomers. Thus, new and original block copolymers that could not be prepared by classical polymerization techniques can be easily obtained using this strategy. The high stability of the *bis*-2,2′:6′,2′terpyridine-ruthenium(II) allows the integrity of block copolymers formed in this way to be kept in various environments, such as organic solvents or water, even under extreme pH (the complex is stable for pH values ranging from 0 to 14) and salt concentrations [131].

In a very recent set of papers [48, 54, 59, 60, 131, 132, 324–328], the synthesis and characterization of metallosupramolecular amphiphilic block copolymers containing a hydrophilic PEO block linked to a hydrophobic PS or PEB block through a *bis*-2,2′:6′,2′terpyridine-ruthenium(II) complex have been described. These copolymers form the so-called "metallosupramolecular micelles".

The *bis*-2,2′:6′,2′terpyridine-ruthenium(II) complexes are assumed to be located at the core/corona interface, as schematically depicted in Fig. 23.

Spherical micelles prepared from PS-[Ru]-PEO copolymers were first investigated by a variety of techniques, including DLS [131, 325], TEM [48], cryo-TEM [54], AFM [325, 326], and AUC [59, 60]. In agreement with previous results obtained on covalent PS-PEO copolymers, PS-[Ru]-PEO micelles had a strong tendency to form clusters [131, 132]. This tendency was even more developed for the metallosupramolecular sample as evidenced from a comparison between a covalent PS_{22}-PEO_{70} and its metallosupramolecular PS_{20}-[Ru]-PEO_{70} counterpart [131]. No clustering was observed for the covalent sample, while it occurred in the metallosupramolecular one. Cryo-

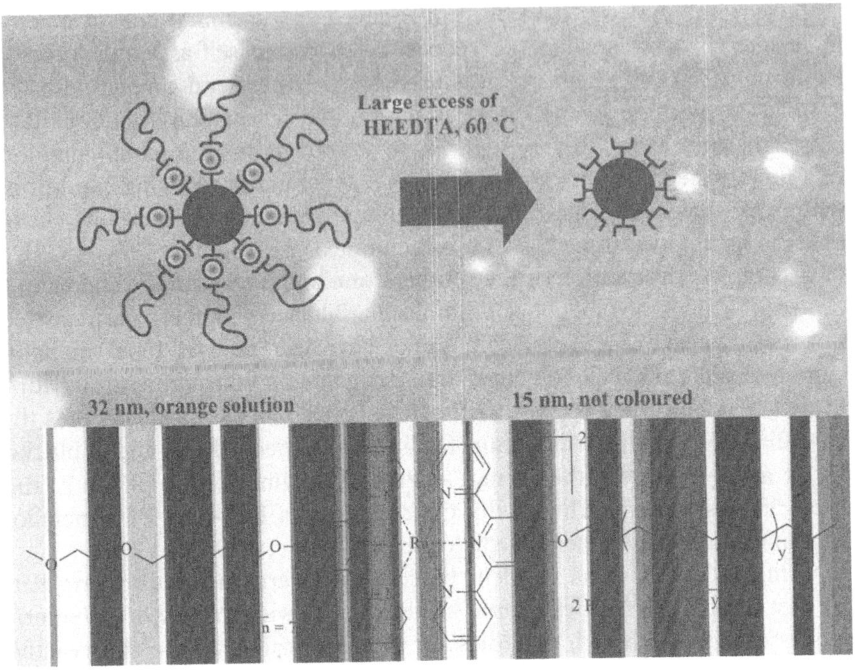

Fig. 23 Schematical representation of metallosupramolecular micelles. *Brackets*: terpyridine ligands; *gray spheres*: Ru^{III} ions. The complexes can be opened by addition of a large excess of hydroxyethyl ethylene diamine tetracetic acid (HEEDTA). This is accompanied by a change in color of micellar solution, which turns from orange to colorless. AFM pictures (*height contrast*) obtained for micelles before and after treatment with HEEDTA are shown in *background*. Average diameter of objects is also indicated. Chemical structure of PEB_{70}-[Ru]-PEO_{70} used for this experiment is shown *below* diagram. Reprinted with permission from [327]. Copyright (2002) Wiley

TEM observations successfully demonstrated that the large aggregates resulted from clustering of individual micelles for the metallosupramolecular micelles [54].

Clustering in PS-[Ru]-PEO micelles was shown to be sensitive to temperature and ionic strength, as previously observed for covalent PS-PEO copolymer [131]. These effects were, however, more pronounced in the case of the metallosupramolecular copolymers, indicating a possible influence of the charged *bis*-2,2′:6′,2′terpyridine-ruthenium(II) complexes [329].

Covalent and metallosupramolecular block copolymers have been combined in a single macromolecular structure. In this respect a terpyridine-functionalized PS-P2VP has been complexed with a terpyridine-functionalized PEO, leading to a PS_{32}-$P2VP_{13}$-[Ru]-PEO_{70} ABC triblock copolymer [330]. This copolymer was further used to prepare core-shell-corona micelles consisting of a PS core, a pH-responsive P2VP shell, and a PEO corona.

Although *bis*-2,2′:6′,2′terpyridine-ruthenium(II) complexes have proven to be extremely stable in various environments, some recent experiments have shown that the addition of a large excess of a competing ligand (hydroxyethyl ethylenediaminetriacetic acid, trisodic salt) did allow the opening of the initial complex [327, 329, 330]. This resulted in the formation of original nano-objects decorated at their surfaces with terpyridine ligands (Fig. 23). This opening was directly seen macroscopically by the disappearance of the characteristic orange color of the *bis*-2,2′:6′,2′terpyridine-ruthenium(II) complex.

In a very recent investigation, hydrophobic PFS (Sect. 7.1) was attached to a hydrophilic PEO block to form an amphiphilic PFS_{12}-[Ru]-PEO_{70} block copolymer [331]. Rodlike micelles were observed in water for this copolymer (Fig. 24). These micelles have a constant diameter but are rather polydisperse in length, and DLS measurements indicate that they are flexible. Crystallization of the PFS in these micelles was observed and is thought to be the key behind the formation of rodlike structures. The cylindrical micelles can be cleaved into smaller rods whenever the temperature of the solution is increased or whenever they are exposed to ultrasound.

Metallosupramolecular graft copolymers have been prepared from a statistical copolymer containing methylmethacrylate and a methacrylate bearing a terpyridine group in the side chain. Terpyridine end-functionalized

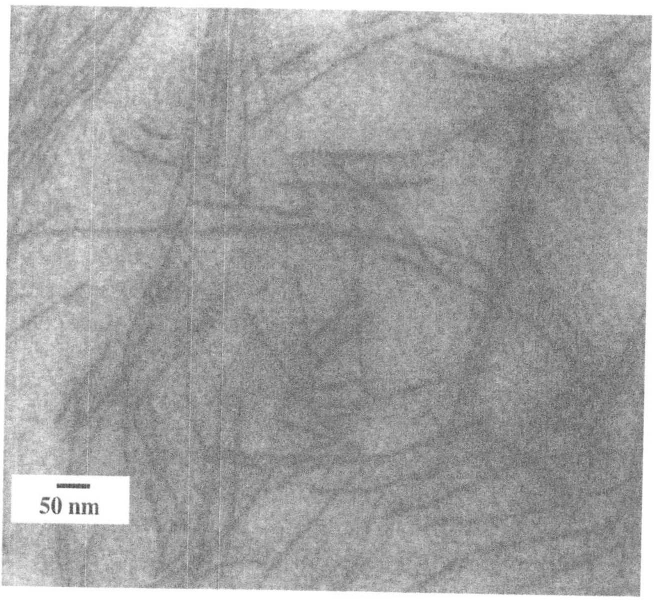

Fig. 24 TEM picture of cylindrical micelles formed by a PFS_{12}-[Ru]-PEO_{70} metallo-supramolecular copolymer in water. No contrasting agent has been added for visualization of these micelles because they contain iron atoms in the core. Reprinted with permission from [331]. Copyright (2004) Wiley

PEO chains have been grafted to this backbone, leading to an amphiphilic copolymer. Aqueous micelles have been then prepared from these copolymers [332].

8
Conclusions and Outlook

Block copolymer micelles have been the subject of an enormous body of work during the last 30 years. Although the basic principles of block copolymer micellization have been already discovered and experimentally investigated in the 80s, intense research on this topic has been performed since the mid-90s by many research groups. Because it was not possible to include every contribution to that field in the frame of the present review, only selected examples were discussed.

The discovery of new "controlled" polymerization techniques in the mid-1990s and the progress achieved in "living" polymerization toward well-defined block copolymers with complex topologies have certainly played a key role in the development of block copolymer micelles.

The interest in block copolymer micelles stems from the numerous practical applications of such systems. These applications were briefly discussed in the framework of some specific block copolymer micelles but were not systematically reviewed here. Practical applications of block copolymer micelles were discussed in the recent review of Riess [14] and include biomedical applications, emulsion stabilization, viscosity regulation, catalyst support, surface modification, etc. Amphiphilic block copolymers with biocompatible and/or biodegradable sequences have potential for applications in the biomedical field, e.g., as drug delivery systems and as carriers in gene therapy. Recent developments in nanotechnology have also created an important demand for self-organized materials including block copolymer micelles. In this respect, block copolymer micelles are used as templating reactors for the production of metallic conducting, semiconducting, or magnetic nanoparticles. In addition to shape and size control due to micellar templating, such nanoparticles embedded in a polymer matrix exhibit enhanced chemical stability to oxidation and better colloidal stability and could also show increased catalytic activity. Block copolymer micelles can also be used as resists for lithographic processes.

Although many investigations have been carried out over the last 30 years on block copolymer micelles, some basic questions have not yet received a clear answer. In this respect, there is a need for libraries of block copolymers in which the relative lengths of the different constituent blocks are systematically varied. Most of the investigations on block copolymer micelles have indeed been realized on a limited number of samples. Synthetic strategies based on the use of macroinitiators or the recent "lego™" approach developed in the frame of metallosupramolecular copolymers could be advantageneously used for the preparation of such series of copolymers.

Although many amphiphilic block copolymer architectures including star-shaped, H-shaped, graft, and ABC block copolymers can now be succesfully synthesized, their micellization behavior remains poorly investigated.

As described in Sect. 7.1, novel "hybrid" block copolymers comprising on one hand a "classical" synthetic polymer and on the other hand a metal-containing polymer, a synthetic or natural (proteine, enzyme, etc.) polypeptide have recently been synthesized. Other "hybrid" block copolymers containing inorganic blocks, dendrimers, etc. will certainly be prepared and thoroughly investigated in the next few years.

The use of noncovalent interactions to trigger micellization from mutually interacting block copolymers is also a topic receiving increasing attention. This has been illustrated for IPEC-containing micelles in Sect. 7.4.

Stimulus-reponsive micelles have been intensively investigated during the last 5 years, and further developments could be expected. These micellar systems have been essentially prepared from double-hydrophilic block copolymers and from micelles with noncovalent complexes in the core. The interest related to these micellar systems stems from their potential applications, as briefly discussed in Sect. 4.3.

In a more general way, the two major driving forces for the design of novel micellar systems are the control over morphology (spheres, vesicles, rods, tubules etc. with controlled size) and function (stimulus-responsive materials, biological functions). Both of these aspects are intimately related since a given morphology can induce a specific function.

Finally, much work remains to be done on concentrated micellar solutions, which have been poorly investigated compared to the dilute regime. Interesting properties, such as lyotropic mesophase behavior, are expected to be observed for these concentrated micellar solutions.

Acknowledgement JFG is grateful to the Communauté française de Belgique for financial support in the frame of the Action de Recherches Concertées NANOMOL.

References

1. Bates FM, Fredrickson GH (1990) Annu Rev Phys Chem 41:525
2. Hamley IW (ed) (1998) The physics of block copolymers. Oxford Science Publications, Oxford
3. Kataoka K, Harada A, Nagasaki Y (2001) Adv Drug Deliv Rev 47:113
4. Price C (1982) Colloidal properties of block copolymers. In: Goodman I (ed) Developments in block copolymers. Applied Science, London
5. Piirma I (ed) (1992) Polymeric surfactants. In: Surfactant science series, vol 42. Marcel Dekker, New York
6. Tuzar Z, Kratochvil P (1993) Micelles of block and graft copolymers in solution. In: Matijevic E (ed) Surface and colloid science, vol 15. Plenum, New York

7. Riess G, Hurtrez G, Bahadur P (1985) Block copolymers, 2nd edn. In: Encyclopedia of polymer science and engineering, vol 2. Wiley, New York
8. Riess G, Dumas PH, Hurtrez G (2002) Block copolymer micelles and assemblies, MML series 5. Citus, London
9. Webber SE, Munk P, Tuzar Z (eds) (1996) Solvents and self-organization of polymers. NATO ASI series, series E: Applied sciences, vol 327. Kluwer, Dordrecht
10. Alexandridis P, Hatton TA (1996) Block copolymers. In: Salamone JC (ed) Polymer materials encyclopedia. CRC Press, Boca Raton, FL
11. Nace VM (ed) (1996) Nonionic surfactants: polyoxyalkylene block copolymers. In: Surfactant science series, vol 60. Marcel Dekker, New York
12. Alexandridis P, Lindman B (eds) (2000) Amphiphilic block copolymers: self assembly and applications. Elsevier, Amsterdam
13. Xie HQ, Xie D (1999) Prog Polym Sci 24:275
14. Riess G (2003) Prog Polym Sci 28:1107
15. Förster S, Abetz V, Müller AHE (2004) Adv Polym Sci 166:173
16. Elias HG (1973) J Macromol Sci Chem 7:601
17. Tsunashima Y, Hirata M, Kawamata Y (1990) Macromolecules 23:1089
18. Alexandridis P, Holzwarth JF, Hatton TA (1994) Macromolecules 27:2414
19. Chu B, Zhou Z (1996) Physical chemistry of polyoxyalkylene block copolymer surfactants. In: Nace VN (ed) Nonionic surfactants: polyoxyalkylene block copolymers, vol 60. Marcel Dekker, New York
20. Wanka G, Hofmann H, Ulbricht W (1994) Macromolecules 27:4145
21. Booth C, Yu GE, Nace VM (1997) Block copolymers of ethylene oxide and 1,2-butylene oxide. In: Alexandridis P, Lindman B (eds) Amphiphilic block copolymers: self-asssembly and applications. Elsevier, Amsterdam
22. Yang YW, Yang Z, Zhou ZK, Atwood D, Booth C (1996) Macromolecules 29:670
23. Zhou Z, Chu B, Nace VM (1996) Langmuir 12:5016
24. Zhou Z, Chu B, Nace VM, Yang YW, Booth C (1996) Macromolecules 29:3663
25. Zhou Z, Yang YW, Booth C, Chu B (1996) Macromolecules 29:8357
26. Yun J, Faust R, Szilagyi LS, Keki S, Zsuga M (2003) Macromolecules 36:1717
27. Pispas S, Hadjichristidis N, Potemkin I, Khokhlov A (2000) Macromolecules 33:1741
28. Zhang L, Eisenberg A (1996) J Am Chem Soc 118:3168
29. Zhou Z, Chu B (1987) Macromolecules 20:3089
30. Zhou Z, Chu B (1988) Macromolecules 21:2548
31. Reddy NK, Fordham PJ, Attwood D, Booth C (1990) J Chem Soc, Faraday Trans 86:1569
32. Munk P (1996) Equilibrium and nonequilibrium polymer micelles. In: Webber SE, Munk P, Tuzar Z (eds) NATO ASI series, series E: Applied sciences, vol 327. Kluwer, Dordrecht
33. Erhardt R, Böker A, Zettl H, Kaya H, Pyckhout-Hintzen W, Krausch G, Abetz V, Müller AHE (2001) Macromolecules 34:1069
34. Guo A, Liu G, Tao J (1996) Macromolecules 29:2487
35. Cameron NS, Corbierre MK, Eisenberg A (1999) Can J Chem 77:1311
36. Zhang L, Eisenberg A (1995) Science 268:1728
37. Zhang L, Shen H, Eisenberg A (1997) Macromolecules 30:1001
38. Yu Y, Zhang L, Eisenberg A (1998) Macromolecules 31:1144
39. Stepanek P (1972) In: Brown W (ed) Dynamic light scattering. Oxford University Press, Oxford
40. Israelachvili JN (ed) (1985) Intermolecular and surface forces. Academic, London
41. Tuzar Z (1996) Overview of polymer micelles. In: Webber SE, Munk P, Tuzar Z (eds) NATO ASI series, series E: Applied sciences, vol 327. Kluwer, Dordrecht

42. Munk P (1996) Classical methods for the study of block copolymer micelles. In: Webber SE, Munk P, Tuzar Z (eds) NATO ASI series, series E: Applied sciences, vol 327. Kluwer, Dordrecht
43. Webber SE (1996) Use of fluorescence methods to characterize the interior of polymer micelles. In: Webber SE, Munk P, Tuzar Z (eds) NATO ASI series, series E: Applied sciences, vol 327. Kluwer, Dordrecht
44. Mortensen K (2000) Small angle scattering studies of block copolymer micelles, micellar mesophases and networks. In: Alexandridis P, Lindman B (eds) Amphiphilic block copolymers: self assembly and applications. Elsevier, Amsterdam
45. Zana R (2000) Fluorescence studies of amphiphilic block copolymers in solution. In: Alexandridis P, Lindman B (eds) Amphiphilic block copolymers: self assembly and applications. Elsevier, Amsterdam
46. Gohy JF, Willet N, Varshney SK, Zhang J-X, Jérôme R (2002) e-Polymers 35
47. Gohy JF, Willet N, Varshney S, Zhang J-X, Jérôme R (2001) Angew Chem Int Ed 40:3214
48. Gohy JF, Lohmeijer BGG, Décamps B, Leroy E, Boileau S, Van Den Broek JA, Schubert D, Haase W, Schubert US (2003) Polym Int 52:1611
49. Esselink FJ, Dormidontova E, Hadziioannou G (1998) Macromolecules 31:2925
50. Lam YM, Grigorieff N, Goldbeck-Wood G (1999) Phys Chem Chem Phys 1:3331
51. Goldraich M, Talmon Y (2000) Direct-imaging cryo-transmission electron microscopy in the study of colloids and polymer solutions. In: Alexandridis P, Lindman B (eds) Amphiphilic block copolymers: self assembly and applications. Elsevier, Amsterdam
52. Mortensen K, Talmon Y, Gao B, Kops J (1997) Macromolecules 30:6764
53. Förster S, Hermsdorf N, Böttcher C, Lindner P (2002) Macromolecules 35:4096
54. Regev O, Gohy JF, Lohmeijer BGG, Varshney SK, Hubert DHW, Schubert US (2004) Colloid Polym Sci 282:407
55. Erhardt R, Zhang M, Böker A, Zettl H, Abetz C, Frederik P, Krausch G, Abetz V, Müller AHE (2003) J Am Chem Soc 125:3260
56. Connell SD, Collins S, Fundin J, Yang Z, Hamley IW (2003) Langmuir 19:10449
57. Hamley IW, Connell SD, Collins S (2004) Macromolecules 37:5337
58. Selb J, Gallot Y (1981) Macromol Chem 182:1513
59. Vogel V, Gohy JF, Lohmeijer BGG, Van Den Broek JA, Haase W, Schubert US, Schubert D (2003) J Polym Sci A 41:3159
60. Mayer G, Vogel V, Lohmeijer BGG, Gohy JF, van den Broek JA, Haase W, Schubert US, Schubert D (2004) J Polym Sci A 42:4458
61. Booth C, Naylor TD, Price C, Rajab NS, Stubbersfield RB (1978) J Chem Soc Faraday Trans I 74:2352
62. Teo HH, Styring MG, Yeates SG, Price C, Booth C (1986) J Coll Interf Sci 114:416
63. Bednar B, Edwards K, Almgren M, Tormod S, Tuzar S (1988) Macromol Chem Rapid Commun 9:785
64. Creutz S, Jérôme R (1999) Langmuir 15:7145
65. Creutz S, Jérôme R (2001) e-Polymers 14
66. Creutz S, van Stam J, Antoun S, De Schryver FC, Jérôme R (1997) Macromolecules 30:4078
67. Creutz S, van Stam J, De Schryver FC, Jérôme R (1998) Macromolecules 31:681
68. van Stam J, Creutz S, de Schryver FC, Jérôme R (2000) Macromolecules 33:6388
69. Riess G, Hurtrez G (1996) Block copolymers: synthesis, colloidal properties and application possibilities of micellar systems. In: Webber SE, Munk P, Tuzar Z (eds) NATO ASI series, series E: Applied sciences, vol 327. Kluwer, Dordrecht

70. Holappa S, Kantonen L, Winnik FM, Tenhu H (2004) Macromolecules 37:7008
71. Won YY, Davis HT, Bates FS (2003) Macromolecules 36:953
72. Stepanek M, Prochazka K (2000) Langmuir 16:2502
73. Prochazka K, Limpouchova Z, Webber SE (1996) Block copolymer micelles: 2. Fluorimetric studies and computer modeling. In: Salamone JC (ed) Polymer materials encyclopedia, vol 1. CRC Press, Boca Raton, FL
74. Farago BJ, Monkenbusch M, Richter D, Huang JS, Fetters LJ, Gast AP (1993) Phys Rev Lett 71:1015
75. Castelletto V, Hamley IW, Yang Z, Haeussler W (2003) J Chem Phys 119:8158
76. Spevacek J (1982) Macromol Chem Rapid Commun 3:697
77. Liu S, Billingham NC, Armes SP (2001) Angew Chem Int Ed 40:2328
78. Gao Z, Zhong XF, Eisenberg A (1994) Macromolecules 27:794
79. Prochazka K, Baloch MK (1979) Macromol Chem 180:2521
80. Tuzar Z, Bednar B, Konak C, Kubin M, Svobodova S, Prochazka K (1982) Macromol Chem 183:399
81. Wilson DJ, Riess G (1988) Eur Polym J 24:617
82. Ding J, Liu G (1998) Chem Mater 10:537
83. Wooley KL (2000) J Polym Sci A 38:1397
84. Zhang Q, Remsen EE, Wooley KL (2000) J Am Chem Soc 122:3642
85. Bütün V, Billingham NC, Armes SP (1999) ACS Polym Prepr Div Polym Chem 40:236
86. Liu T, Liu LZ, Chu B (2000) Formation of amphiphilic block copolymer micelles in nonaqueous solution. In: Alexandridis P, Lindman B (eds) Amphiphilic block copolymers: self-assembly and applications. Elsevier, Amsterdam
87. Pitsikalis M, Siakali-Kioulafa E, Hadjichristidis N (2000) Macromolecules 33:5460
88. Quintana JR, Salazar RA, Katime I (1995) Macromol Chem Phys 196:1625
89. Quintana JR, Janez MD, Katime I (1995) Macromol Rapid Commun 16:607
90. McClain JB, Betts DE, Canelas DA, Samulski ET, DeSimone JM, Londono JD, Cochran HD, Wignall GD, Chillura-Martino D, Triolo R (1996) Science 274:2049
91. Raudino A, Lo Celso F, Triolo A, Triolo R (2004) J Chem Phys 120:3489
92. Samii AA, Lindman B, Karlstrom G (1990) Prog Colloid Polym Sci 82:280
93. Samii AA, Karlstrom G, Lindman B (1991) Langmuir 7:1067
94. Alexandridis P, Yang L (2000) Macromolecules 33:3382
95. Gervais M, Gallot B (1977) Macromol Chem 178:2071
96. Gast AP, Vinson PK, Cogan-Farinas KA (1993) Macromolecules 26:1774
97. Lin EK, Gast AP (1996) Macromolecules 29:4432
98. Vagberg LJM, Cogan KA, Gast AP (1991) Macromolecules 24:1670
99. Wu G, Chu B (1994) Macromolecules 27:1766
100. Guo C, Liu HZ, Chen JY (2000) Colloids Surf A 175:193
101. Kovacs AJ, Manson JA (1966) Kolloid Z Z Polym 214:1
102. Spatz JP, Roescher A, Möller M (1996) Adv Mater 8:337
103. Bronstein LM, Chernyshov DM, Timofeeva GI, Dubrovina LV, Valetsky PM, Obolonkova ES, Khokhlov AR (2000) Langmuir 16:3626
104. Moffitt M, Zhang L, Khougaz K, Eisenberg A (1996) Micellization of ionic block copolymers in three dimensions. In: Webber SE, Munk P, Tuzar Z (eds) NATO ASI series, series E: Applied sciences, vol 327. Kluwer, Dordrecht
105. Zhang L, Khougaz K, Moffitt M, Eisenberg A (2000) Self assembly of block polyelectrolytes. In: Alexandridis P, Lindman B (eds) Amphiphilic block copolymers: self-assembly and applications. Elsevier, Amsterdam
106. Förster S, Zisenis M, Wenz E, Antonietti M (1996) J Chem Phys 104:9956
107. Förster S, Antonietti M (1998) Adv Mater 10:195

108. Klingelhofer S, Heitz W, Greiner A, Oestreich S, Förster S, Antonietti M (1997) J Am Chem Soc 119:10116
109. Spatz JP, Röscher A, Sheiko S, Krausch G, Möller M (1995) Adv Mater 7:73
110. Spatz JP, Sheiko S, Möller M (1996) Macromolecules 29:3220
111. Roescher A, Möller M (1995) Adv Mater 7:151
112. Antonietti M, Wenz E, Bronstein LM, Seregina MS (1995) Adv Mater 7:1000
113. Sulman E, Bodrova Y, Matveeva V, Semagina N, Cerveny L, Kurtc V, Bronstein L, Platonova O, Valetsky P (1999) Appl Catal A 176:75
114. Spatz JP, Mössmer S, Möller M (1996) Angew Chem Int Ed 35:1510
115. Mössmer S, Spatz JP, Möller M, Aberle T, Schmidt J, Burchard W (2000) Macromolecules 33:4791
116. Kiriy A, Gorodyska G, Minko S, Stamm M, Tsitsilianis C (2003) Macromolecules 36:8704
117. Almgren M, Brown W, Hvidt S (1995) Colloid Polym Sci 273:2
118. Booth C, Attwood D (2000) Macromol Rapid Commun 21:501
119. Wanka G, Hoffmann H, Ulbricht W (1990) Colloid Polym Sci 268:101
120. Mortensen K, Pedersen JS (1993) Macromolecules 26:805
121. Jain NJ, Aswal VK, Goyal PS, Bahadur P (2000) Colloids Surf A 173:85
122. Guo C, Wang J, Liu HZ, Chen JY (1999) Langmuir 15:2703
123. Chaibundit C, Mai SM, Heatley F, Booth C (2000) Langmuir 16:9645
124. Riess G, Rogez D (1982) ACS Polym Prepr (Div Polym Chem) 23:19
125. Jada A, Hurtrez G, Siffert B, Riess G (1996) Macromol Chem Phys 197:3697
126. Hurtrez G, Dumas Ph, Riess G (1998) Polym Bull 40:203
127. Khan TN, Mobbs RH, Price C, Quintana JR, Stubbersfield RB (1987) Eur Polym J 23:191
128. Xu R, Winnik MA, Riess G, Chu B, Croucher MD (1992) Macromolecules 25:644
129. Mortensen K, Brown W, Almdal K, Alami E, Jada A (1997) Langmuir 13:3635
130. Bronstein LM, Chernyshov DM, Timofeeva GI, Dubrovina LV, Valetsky PM, Khokhlov AR (1999) Langmuir 15:6195
131. Gohy JF, Lohmeijer BGG, Varshney SK, Schubert US (2002) Macromolecules 35:7427
132. Gohy JF, Lohmeijer BGG, Schubert US (2004) ACS Polym Prepr Div Polym Chem 45:395
133. Polik WF, Burchard W (1983) Macromolecules 16:978
134. Yu K, Eisenberg A (1996) Macromolecules 29:6359
135. Rolland A, O'Mullane JE, Goddard P, Brookman L, Petrak K (1992) J Appl Polym Sci 44:1195
136. Cammas S, Kataoka K (1995) Macromol Chem Phys 196:1899
137. Jarrett P, Lalor CB, Chan L, Redmon MP, Hickey AJ (2000) Colloids Surf B 17:11
138. Luo L, Tam D, Maysinger D, Eisenberg A (2002) Bioconjug Chem 13:1259
139. Savic R, Luo L, Eisenberg A, Maysinger D (2003) Science 300:615
140. Nuopponen M, Ojala J, Tenhu H (2004) Polymer 45:3643
141. Verdonck B, Goethals EJ, Du Prez FE (2003) Macromol Chem Phys 204:2090
142. Selb J, Gallot Y (1985) In: Goodman I (ed) Developments in block copolymers, 2nd Edn. Elsevier, Amsterdam
143. Selb J, Gallot Y (1980) Macromol Chem 181:809
144. Astafieva I, Zhong XF, Eisenberg A (1993) Macromolecules 26:7339
145. Astafieva I, Khougaz K, Zhong XF, Eisenberg A (1995) Macromolecules 28:7127
146. Zhang LF, Barlow RJ, Eisenberg A (1995) Macromolecules 28:6055
147. Khougaz K, Astafieva I, Zhong XF, Eisenberg A (1995) Macromolecules 28:7135
148. Tuzar Z (1996) Copolymer micelles in aqueous media. In: Webber SE, Munk P, Tuzar Z (eds) NATO ASI series, series E: Applied sciences, vol 327. Kluwer, Dordrecht

149. Zhang L, Eisenberg A (1996) Science 272:1777
150. Zhang L, Eisenberg A (1996) Macromolecules 29:8805
151. Eisenberg A (2000) ACS Polym Prepr (Div Polym Chem) 41:1515
152. Guenoun P, Davis HT, Tirrell M, Mays JW (1996) Macromolecules 29:3965
153. Guenoun P, Muller F, Delsanti M, Auvray L, Chen YJ, Mays JW, Tirrell M (1998) Phys Rev Lett 81:3872
154. Förster S, Hermsdorf N, Böttcher C, Lindner P (2002) Macromolecules 35:4096
155. Förster S, Hermsdorf N, Leube W, Schnablegger H, Regenbrecht M, Akari S, Lindner P, Böttcher C (1999) J Phys Chem B 103:6652
156. Gohy JF, Antoun S, Jérôme R (2001) Polymer 42:8637
157. Raviv U, Giasson S, Kampf N, Gohy JF, Jérôme R, Klein J (2003) Nature 425:163
158. Vamvakaki M, Billingham NC, Armes SP (1998) Polymer 39:2331
159. Bütün V, Vamvakaki M, Billingham NC, Armes SP (2000) Polymer 41:3173
160. Liu S, Armes SP (2001) Curr Opin Colloid Interf Sci 6:249
161. Unali GF, Armes SP, Billingham NC, Tuzar Z, Hamley IW (1999) ACS Polym Prepr (Div Polym Chem) 40:259
162. Wendler U, Bohrisch J, Jaeger W, Rother G, Dautzenberg H (1998) Macromol Rapid Commun 19:185
163. Leyh B, Creutz S, Gaspard JP, Bourgaux C, Jérôme R (1998) Macromolecules 31:9258
164. Cölfen H (2001) Macromol Rapid Commun 22:219
165. Martin TJ, Prochazka K, Munk P, Webber SE (1996) Macromolecules 29:6071
166. Gohy JF, Mores S, Varshney SK, Zhang JX, Jérôme R (2002) e-polymers 21
167. Bronstein LM, Sidorov SN, Valetsky PM, Hartmann J, Cölfen H, Antonietti M (1999) Langmuir 15:6256
168. Sidorov SN, Bronstein LM, Kabachii YA, Valetsky PM, Soo PL, Maysinger D, Eisenberg A (2004) Langmuir 20:3543
169. Mathur AM, Drescher B, Scranton AB, Klier J (1998) Nature 392:367
170. Gohy JF, Varshney SK, Jérôme R (2001) Macromolecules 34:3361
171. Holappa S, Karesoja M, Shan J, Tenhu H (2002) Macromolecules 35:4733
172. Dou H, Jiang M, Peng H, Chen D, Hong Y (2003) Angew Chem Int Ed 42:1516
173. Kamachi M, Kurihara M, Stille JK (1972) Macromolecules 5:161
174. Gohy JF, Creutz S, Garcia M, Mahltig B, Stamm M, Jérôme R (2000) Macromolecules 33:6378
175. Golub T, de Keizer A, Cohen-Stuart MA (2000) Macromolecules 33:767
176. Lowe AB, Billingham NC, Armes SP (1997) Chem Commun 1035
177. Mahltig B, Gohy JF, Jérôme R, Bellmann C, Stamm M (2000) Colloid Polym Sci 278:502
178. Mahltig B, Gohy JF, Jérôme R, Stamm M (2001) J Polym Sci B 39:709
179. Mahltig B, Müller-Bushbaum P, Wolkenhauer M, Wunnicke O, Wiegand S, Gohy JF, Jérôme R, Stamm M (2001) J Coll Interf Sci 242:36
180. Mahltig B, Gohy JF, Jérôme R, Buchhammer HM, Stamm M (2002) J Polym Sci part B 40:338
181. Varoqui R, Tran Q, Pefferkorn E (1979) Macromolecules 12:831
182. Bekturov EA, Kudaibergenov SE, Khamzamulina RE, Frolova VA, Nurgalieva DE, Schulz RC, Zöller J (1992) Macromol Chem Rapid Commun 13:225
183. Bekturov EA, Kudaibergenov SE, Frolova VA, Khamzamulina RE, Schulz RC, Zöller J (1991) Macromol Chem Rapid Commun 12:37
184. Gabaston LI, Furlong SA, Jackson RA, Armes SP (1999) Polymer 40:4505
185. Bütün V, Billingham NC, Armes SP (1997) Chem Commun 671
186. Gohy JF, Antoun S, Jérôme R (2001) Macromolecules 24:7435

187. Mahltig B, Gohy JF, Antoun S, Jérôme R, Stamm M (2002) Colloid Polym Sci 280:495
188. Bütün V, Bennett CE, Vamvakaki M, Lowe AB, Billingham NC, Armes SP (1997) J Mater Chem 7:1693
189. Nishi T, Kwei K (1975) Polymer 16:285
190. Aoshima S, Kobayashi E (1995) Makromol Chem Makromol Symp 95:91
191. Topp MDC, Dijkstra PJ, Talsma H, Feijen J (1997) Macromolecules 30:8518
192. Forder C, Patrickios CS, Billingham NC, Armes SP (1996) Chem Commun 883
193. Forder C, Patrickios CS, Armes SP, Billingham NC (1996) Macromolecules 29:8160
194. Arotçarena M, Heise B, Ishaya S, Laschewsky A (2002) J Am Chem Soc 124:3787
195. Gast AP (1996) Structure and interaction in tethered chains. In: Webber SE, Munk P, Tuzar Z (eds) NATO ASI series, series E: Applied sciences, vol 327. Kluwer, Dordrecht
196. Linse P (2000) Modelling of self-assembly of block copolymers in selective solvent. In: Alexandridis P, Lindman B (eds) Amphiphilic block copolymers: self assembly and applications. Elsevier, Amsterdam
197. Noolandi J, Hong KM (1983) Macromolecules 16:1443
198. Leibler L, Orland H, Wheeler JC (1983) J Chem Phys 79:3550
199. Nagarajan R, Ganesh K (1989) Macromolecules 22:4312
200. Hurter PN, Scheutjens JMHM, Hatton TA (1993) Macromolecules 26:5592
201. Binder K (ed) (1995) Monte Carlo and molecular dynamics simulations in polymer Science. Oxford University Press, New York
202. Binder K, Müller M (2000) Curr Opin Colloid Interf Sci 5:315
203. Haliloglu T, Mattice WL (1996) Monte Carlo simulations of self-assembly in macromolecular systems. In: Webber SE, Munk P, Tuzar Z (eds) NATO ASI series, series E: Applied sciences, vol 327. Kluwer, Dordrecht
204. Marko JF, Rabin Y (1992) Macromolecules 25:1503
205. Dan N, Tirrell M (1993) Macromolecules 26:4310
206. Shusharina NP, Nyrkova IA, Khokhlov AR (1996) Macromolecules 29:3167
207. Daoud M, Cotton JP (1982) J Phys 43:531
208. Halperin A (1987) Macromolecules 20:2943
209. Zhulina EB, Birshtein TM (1986) Polym Sci USSR 27:570
210. Shelley JC, Shelley MY (2000) Curr Opin Colloid Interf Sci 5:101
211. Förster S, Plantenberg T (2002) Angew Chem Int Ed 41:688
212. Qin A, Tian M, Ramireddy C, Webber SE, Munk P, Tuzar Z (1994) Macromolecules 27:120
213. Ahrens H, Förster S, Helm CA (1998) Phys Rev Lett 81:4172
214. Förster S, Hermsdorf N, Leube W, Schnablegger H, Regenbrecht M, Akari S, Lindner P, Böttcher C (1999) J Phys Chem B 103:6652
215. Eckert AR, Webber SE (1996) Macromolecules 29:560
216. Voulgaris D, Tsitsilianis C, Grayer V, Esselink E, Hadziioannou G (1990) Polymer 40:5879
217. Borisov OV, Zhulina EB (2002) Macromolecules 35:4472
218. Guenoun P, Schlachli A, Sentenac D, Mays JW, Benattar JJ (1995) Phys Rev Lett 74:3628
219. Hariharan R, Biver C, Mays JW, Russell WB (1998) Macromolecules 31:7506
220. Hariharan R, Biver C, Russell WB (1998) Macromolecules 31:7514
221. Tauer K, Müller H, Rosengarten L, Riedelsberger K (1999) Colloids Surfaces A Physicochem Eng Aspects 153:75
222. Luo L, Eisenberg A (2001) J Am Chem Soc 123:1012
223. Luo L, Eisenberg A (2001) Langmuir 17:6804
224. Dobereiner HG, Evans E, Kraus M, Seifert U, Wortis M (1997) Phys Rev E 55:4458

225. Luo L, Eisenberg A (2002) Angew Chem Int Ed 41:1001
226. Antonietti M, Heinz S, Schmidt M, Rosenauer C (1994) Macromolecules 27:3276
227. Tao J, Stewart S, Liu G, Yang M (1997) Macromolecules 30:2738
228. Ding J, Liu G (1997) Macromolecules 30:655
229. Won YY, Davis HT, Bates FS (1999) Science 283:960
230. Discher BM, Won YY, Ege DS, Lee JCM, Bates FS, Discher DE, Hammer DA (1999) Science 284:1143
231. Aranda-Spinoza H, Bermudez H, Bates FS, Discher DE (2001) Phys Rev Lett 87:208301
232. Lee JCM, Santore M, Bates FS, Discher DE (2002) Macromolecules 35:323
233. Dimova R, Seifert B, Pouligny D, Forster S, Döbereiner HG (2002) Eur Phys J E 7:241
234. Schillen K, Bryskhe K, Melnikova YS (1999) Macromolecules 32:6885
235. Nardin C, Hirt T, Leukel J, Meier W (2000) Langmuir 16:1035
236. Grumelard J, Taubert A, Meier W (2004) Chem Commun 1462
237. Sommerdijk NAJM, Holder SJ, Hiorns RC, Jones RG, Nolte RJM (2000) Macromolecules 33:8289
238. Cornelissen JJLM, Rowan AE, Nolte RJM, Sommerdijk NAJM (2001) Chem Rev 101:4039
239. Discher DE, Eisenberg A (2002) Science 297:967
240. Lee M, Cho BK, Zin WC (2001) Chem Rev 101:3869
241. Klok HA, Lecommandoux S (2001) Adv Mater 13:1217
242. Bates FS, Schulz MF, Rosedale JH, Almdal K (1992) Macromolecules 25:5547
243. Chen JT, Thomas EL, Ober CK, Hwang SS (1995) Macromolecules 28:1688
244. Chen JT, Thomas EL, Ober CK, Mao GP (1996) Science 273:343
245. Jenekhe, Chen XL (1999) Science 283:372
246. Chen XL, Jenekhe SA (1999) Langmuir 15:8007
247. Stupp SI, LeBonheur V, Walker K, Li LS, Huggins KE, Keser M, Amstutz A (1997) Science 276:384
248. Cornelissen JJLM, Fischer M, Sommerdijk NAJM, Nolte RJM (1998) Science 280:1427
249. Leclère P, Calderone A, Marsitzky D, Francke V, Geerts Y, Müllen K, Brédas JL, Lazzaronni R (2000) Adv Mater 12:1042
250. Reiter G, Hoerner P, Hurtrez G, Riess G, Sommer JU, Joanny JF (1998) J Surf Sci Technol 14:93
251. Massey JA, Power KN, Manners I, Winnik MA (1998) J Am Chem Soc 120:9533
252. Djalali R, Hugenberg N, Fischer K, Schmidt M (1999) Macromol Rapid Commun 20:444
253. Massey JA, Temple K, Cao L, Rharbi Y, Raez J, Winnik MA, Manners I (2000) J Am Chem Soc 122:11577
254. Massey JA, Winnik MA, Manners I, Chan VZH, Ostermann JM, Enchelmaier R, Spatz JP, Möller M (2001) J Am Chem Soc 123:3147
255. Cao L, Massey JA, Winnik MA, Manners I, Riethmüller S, Banhart F, Möller M (2003) Adv Funct Mater 13:271
256. Wang XS, Winnik MA, Manners I (2002) Macromol Rapid Commun 23:210
257. Raez J, Barjonovu R, Massey JA, Winnik MA, Manners I (2000) Angew Chem Int Ed 39:3862
258. Cao L, Manners I, Winnik MA (2001) Macromolecules 34:3353
259. Resendes R, Massey JA, Temple K, Cao L, Power-Billard KN, Winnik MA, Manners I (2001) Chem Eur J 7:2414
260. Resendes R, Massey JA, Dorn H, Winnik MA, Manners I (2000) Macromolecules 33:8
261. Chécot F, Lecommandoux S, Gnagnou Y, Klok HA (2002) Angew Chem Int Ed 41:1340

262. Klok HA (2002) Angew Chem Int Ed 41:1509
263. Crespo JS, Lecommandoux S, Borsali R, Klok HA, Soldi V (2003) Macromolecules 36:1253
264. Velonia K, Rowan AE, Nolte RJM (2002) J Am Chem Soc 124:4224
265. Breiner U, Krappe U, Abetz V, Stadler R (1997) Macromol Chem Phys 198:1051
266. Kriz J, Masar B, Plestil J, Tuzar Z, Pospisil H, Doskocilova D (1998) Macromolecules 31:41
267. Ishizone T, Sugiyama K, Sakano Y, Mori H, Hirao A, Nakahama S (1999) Polymer J 31:983
268. Prochazka K, Martin TJ, Webber SE, Munk P (1996) Macromolecules 29:6526
269. Tsitsilianis C, Voulgaris D, Stepanek M, Podhajecka K, Prochazka K, Tuzar Z, Brown W (2000) Langmuir 16:6868
270. Yu G, Eisenberg A (1998) Macromolecules 31:5546
271. Lei L, Gohy JF, Willet N, Zhang JX, Varshney S, Jérôme R (2004) Macromolecules 37:1089
272. Lei L, Gohy JF, Willet N, Zhang JX, Varshney S, Jérôme R (2004) Polymer 45:4375
273. Stepanek M, Humpolickova J, Prochakza K, Hof M, Tuzar Z, Spirkova M, Wolff T (2003) Collect Czech Chem Commun 68:121
274. Khanal A, Li Y, Takisawa N, Kawasaki N, Oishi Y, Nakashima K (2004) Langmuir 20:4809
275. Liu S, Weaver JVM, Tang Y, Billingham NC, Armes SP, Tribe K (2002) Macromolecules 35:6121
276. Giebeler E, Stadler R (1997) Macromol Chem Phys 198:3815
277. Patrickios CS, Forder C, Armes SP, Billingham NC (1997) J Polym Sci Part A 35:1181
278. Ma Q, Wooley KL (2000) J Polym Sci A 38:4805
279. Pochan DJ, Chen Z, Cui H, Hales K, Qi K, Wooley KL (2004) Science 306:94
280. Liu G, Stewart S (1999) Polym Mater Sci Eng 81:10
281. Underhill RS, Liu G (2000) Chem Mater 12:2082
282. de Paz Banez MV, Robinson KL, Bütün V, Armes SP (2000) Polymer 42:29
283. Lambert O, Reutenauer S, Hurtrez G, Dumas P (2000) Macromolecular Symposium, Weinheim vol 161. Wiley, New York
284. Li Z, Kesselman E, Talmon Y, Hillmyer MA, Lodge TP (2004) Science 306:98
285. Patrickios CS, Hertler WR, Abbott NL, Hatton TA (1994) Macromolecules 27:930
286. Chen W-Y, Alexandridis P, Su C-K, Patrickios CS, Hertler WR, Hatton TA (1995) Macromolecules 28:8604
287. Patrickios CS, Lowe AB, Armes SP, Billingham NC (1998) J Polym Sci A 36:617
288. Tsitsilianis C, Sfika V (2001) Macromol Rapid Commun 22:647
289. Bieringer R, Abetz V, Müller AHE (2001) Eur Phys J E Soft Matter 5:5
290. Goldacker T, Abetz V, Stadler R, Erukhimovich I, Leibler L (1999) Nature 398:137
291. Tew GN, Li L, Stupp SI (1998) J Am Chem Soc 120:5601
292. Pralle MU, Urayama K, Tew GN, Neher D, Wegner G, Stupp SI (2000) Angew Chem Int Ed 39:1486
293. Erhardt R, Böker A, Zettl H, Kaya H, Pyckhout-Hintzen W, Krausch G, Abetz V, Müller AHE (2001) Macromolecules 34:1069
294. Xu H, Erhardt R, Abetz V, Müller AHE, Goedel EA (2001) Langmuir 17:678
295. Saito R, Fujita A, Ichimura A, Ishizu K (2000) J Polym Sci A 38:2091
296. Kabanov AV, Bronich TK, Kabanov VA, Yu K, Eisenberg A (1996) Macromolecules 29:6797
297. Gohy JF, Varshney SK, Antoun S, Jérôme R (2000) Macromolecules 33:9298
298. Gohy JF, Varshney SK, Jérôme R (2001) Macromolecules 24:2745

299. Harada A, Kataoka K (1998) Macromolecules 31:288
300. Harada A, Kataoka K (1999) Langmuir 15:4208
301. Harada A, Kataoka K (2001) J Controlled Release 2:85
302. Harada A, Kataoka K (1999) Science 283:65
303. De Smedt SC, Demeester J, Hennink WE (2000) Pharm Res 17:113
304. Wolfert MA, Schacht EH, Toncheva V, Ulbrich K, Nazarova O, Seymour LW (1996) Hum Gene Ther 7:2123
305. Katayose S, Kataoka K (1997) Bioconjug Chem 8:702
306. Harada A, Kataoka K (1995) Macromolecules 28:5294
307. Kataoka K, Harada A, Wakebayashi D, Nagasaki Y (1999) Macromolecules 32:6892
308. Nagasaki Y, Wakebayashi D, Akiyama Y, Harada A, Kataoka K (2000) ACS Polym Prepr (Div Polym Chem) 41:1649
309. Caputto A, Betti M, Altavilla G, Bonaccorsi A, Boarini C, Marchisio M, Butto S, Sparnacci K, Laus M, Tondelli L, Ensoli B (2002) Vaccine 20:2303
310. Liu S, Zhu H, Zhao H, Jiang M, Wu C (2000) Langmuir 16:3712
311. Schrage S, Sigel R, Schlaad H (2003) Macromolecules 36:1417
312. Gohy JF, Khousakhoun E, Willet N, Varshney SK, Jérôme R (2004) Macromol Rapid Commun 25:1536
313. Ruokolainen R, Mäkinen R, Torkelli M, Mäkelä T, Serimaa R, ten Brinke G, Ikkala O (1998) Science 280:557
314. Zhou S, Chu B (2000) Adv Mater 12:545
315. Kabanov AV, Bronich TK, Kabanov VA, Yu K, Eisenberg A (1998) J Am Chem Soc 120:9941
316. Bronich TK, Popov AM, Eisenberg A, Kabanov VA, Kabanov AV (2000) Langmuir 16:481
317. Bronich TK, Ouyang M, Kabanov VA, Eisenberg A, Szoka FC, Kabanov AV (2002) J Am Chem Soc 124:11872
318. Gohy JF, Mores S, Varshney SK, Jérôme R (2003) Macromolecules 36:2579
319. Peng H, Chen D, Jiang M (2003) J Phys Chem B 107:12461
320. Peng H, Chen D, Jiang M (2003) Langmuir 19:10989
321. Constable EC (1986) Adv Inorg Chem Radiochem 30:69
322. Hogg R, Wilkins RG (1962) J Chem Soc 341:350
323. Sullivan BP, Calvert J, Meyer TJ (1980) Inorg Chem 19:1404
324. Lohmeijer BGG, Schubert US (2002) Angew Chem Int Ed 41:3825
325. Gohy JF, Lohmeijer BGG, Schubert US (2002) Macromolecules 35:4560
326. Gohy JF, Lohmeijer BGG, Schubert US (2002) ACS Polym Prepr (Div Polym Chem) 43:364
327. Gohy JF, Lohmeijer BGG, Schubert US (2002) Macromol Rapid Commun 23:555
328. Lohmeijer BGG, Gohy JF, Schubert US (2003) ACS Polym Prepr (Div Polym Chem) 44:585
329. Gohy JF, Lohmeijer BGG, Schubert US (2003) Chem Eur J 9:3472
330. Gohy JF, Lohmeijer BGG, Varshney SK, Décamps B, Leroy E, Boileau S, Schubert US (2002) Macromolecules 35:9748
331. Gohy JF, Lohmeijer BGG, Alexeev A, Wang XS, Manners I, Winnik MA, Schubert US (2004) Chem Eur J 10:4315
332. Gohy JF, Hofmeier H, Alexeev A, Schubert US (2003) Macromol Chem Phys 204:1524

Adv Polym Sci (2005) 190: 137–181
DOI 10.1007/12_002

Published online: 23 September 2005

Nanoporous Materials from Block Copolymer Precursors

Marc A. Hillmyer

Department of Chemistry, University of Minnesota, 207 Pleasant St. SE,
Minneapolis, MN 55455-0431, USA
hillmyer@chem.umn.edu

Abstract Block copolymers constitute a fascinating set of self-assembled materials exhibiting compositional heterogeneities on the nanometer length scale. While traditionally employed as thermoplastic elastomers, asphalt modifiers, and adhesives, the potential of self-assembled block copolymers for nanotechnological applications has been realized in the past decade and many examples have now appeared in the literature. As indicated by the title, this review covers a specific aspect of block copolymers as tools for preparing nanoscopic materials. Nanoporous materials can be generated by selective removal of one component from a self-assembled block copolymer. These materials exhibit the pore size and pore topology of their parent structures and can be used as nanolithographic masks, separation membranes and nanomaterial templates. The work described within covers the published work in the field since the first report of nanoporous materials from ordered block copolymers nearly two decades ago. After an introductory section and historical account, sections on nanolithography, membranes, monoliths, and templates follow. The review ends with a summary and outlook on this exciting research arena in block copolymer science and technology.

Keywords Block copolymers · Etching · Nanolithography · Nanoporous · Self-assembly

Abbreviations

AFM	atomic force microscopy
DBS	dodecylbenzenesulfonate
BET	Brunauer–Emmett–Teller
χ	Flory–Huggins interaction parameter

GASAXS	Grazing angle small-angle X-ray scattering
HABA	2-(4′-hydroxybenzene-azo)benzoic acid
IR	infrared
NMR	nuclear magnetic resonance
PαMS	poly(α-methyl styrene)
PBD	polybutadiene
PCEMA	poly(2-cinnamoylethyl methacrylate)
PCHE	polycyclohexylethylene
PDP	pentadecyl phenol
PDMS	polydimethylsiloxane
PEO	poly(ethylene oxide)
PFMA	poly(perfluorooctylethyl methacrylate)
PHOST	poly(4-hydroxy styrene)
PI	polyisoprene
PLA	polylactide
PLLA	poly-*L*-lactide
PMMA	poly(methyl methacrylate)
PPO	poly(propylene oxide)
P(PMDSS)	poly(pentamethyldisilylstyrene)
PPS	poly(4-vinylphenyl-dimethyl-2-propoxysilane)
PS	polystyrene
PVP	poly-4-vinylpyridine
PtBA	poly(*t*-butylacrylate)
RIE	reactive ion etching
SANS	small-angle neutron scattering
SAXS	small-angle X-ray scattering
$scCO_2$	super critical carbon dioxide
SEC	size exclusion chromatography
SEM	scanning electron microscopy
TBAF	tetrabutyl ammonium fluoride
TEM	transmission electron microscopy
T_g	glass transition temperature
THF	tetrahydrofuran
UV	ultraviolet
WAXS	wide-angle X-ray scattering
XPS	X-ray photoelectron spectroscopy

1 Introduction

Covalently connecting two incompatible polymers at their ends leads to a fascinating class of self-assembling materials [1]. Block copolymers constitute a well-studied and well-documented set of nanostructured hybrid materials [2]. Many synthetic techniques are available for generating AB diblock, ABA triblock, ABC triblock and even more complicated block architectures [3, 4]. Furthermore, the thermodynamics governing the self-assembly of AB diblock copolymers is now established from both experimental [5] and theoretical [6] perspectives. In linear AB diblock copolymers the following four equilibrium morphologies have been identified in numerous systems: lamellar, hexagonally packed cylindrical, bicontinuous gyroid, and BBC spherical. In fact, these four morphologies have been observed in block copolymers of quite different chemical make-up [2]. The phase behavior of coil-coil AB diblocks is predictable; if the tails of two incompatible polymers are covalently connected, one can expect self-assembly into one of four equilibrium symmetries. The composition of the material will dictate the identity of the dispersed and continuous phase(s). In nearly all cases, the principal spacings in these ordered structures are between 5 and 50 nm, thus highlighting the potential utility of these materials in nanotechnological applications [7].

This experimental phenomenology and theoretical underpinning suggests that the self-assembly of AB diblock copolymers is in fact universal, and materials with discrete functionality in the blocks can be expected to adopt one of these four morphologies provided the materials are of appropriate molecular weight, incompatibility, and composition. As a result the study of functional block copolymers that form ordered materials with compositional heterogeneities on the nanometer length scale has been quite active over the past decade. Simply put, the combination of exquisite structural control with some higher-order functionality is on the forefront of block copolymer science and technology. For example, block copolymers containing conducting materials [8], hydrogen-bonding moieties [9], ligands [10], metal-containing segments [11], and degradable components have all been explored with various applications in mind. The latter set of materials is the focus of this review. By preparing block copolymers in which the minority component is degradable (etchable), ordered block copolymers provide a rather convenient route to nanoporous organic materials. For example, selective etching of the cylinder-forming phase in an ordered block copolymer results in the formation of nanoscopic channels provided the matrix material can support such a structure. As will be discussed in detail below, block copolymers are quite versatile materials in this regard; the structures of the resultant nanoporous materials can be precisely tailored.

Nanoporous materials have been shown to be useful for many applications including separations, catalysis, and templating. Nanoporous materials come in all varieties with newly developed inorganic metal oxides being a prevalent example [12]. While these materials have established utility, organic nanoporous materials can find applications where their inorganic cousins do not perform as well. Porous organic polymers are widely used in filtration and chromatographic media. This is due in part to their facile processing characteristics and synthetic accessibility. In the case of nanoporous materials from block copolymer precursors, one can envision many possible applications. These include high surface area supports, nanomaterial templates, and size-specific separation media. Given the range of block copolymer structures available, nanoporous materials with a wide range of tunable properties can be imagined. There are two key requirements for preparing nanoporous materials from ordered block copolymers: (i) the etchable material must be physically accessible to the solvent, reagent, process utilized for degradation; and (ii) the matrix material must be able to support the resultant nanoporous structure. As will be reviewed, there are many possible degradation techniques/etchable blocks and a variety of matrix materials that meet these necessities. As synthetic techniques keep expanding the range of available block copolymer structures, this list will expand, and new applications will parallel this growth.

In this review I have attempted to cover all the reported work on the preparation of nanoporous materials from ordered block copolymer precursors starting in 1988. To the best of my knowledge there are no prior published accounts in the research area. This review begins with an historical account of important early works. This is followed by a section on block copolymer nanolithography, arguably the most developed use of nanoporous materials from ordered block copolymers. The next two sections cover the generation of nanoporous membranes and monoliths. A section covering the use of these nanoporous materials for templating and related applications follows. A short summary and outlook on this exciting field is presented at the end of this review.

2 Historical Developments

The first demonstrated example of generating a porous membrane from an ordered block copolymer was communicated by Lee et al. in 1988 [13]. In that work, a triblock copolymer of poly(4-vinylphenyl-dimethyl-2-propoxysilane)-*b*-polyisoprene-*b*-poly(4-vinylphenyl-dimethyl-2-propoxysilane) [PPS-PI-PPS] that formed a lamellar morphology was cast into a thin (ca. 20 μm) film (no effort was made to orient the block copolymer domains, although some orientation presumably resulted from the casting procedure), crosslinked by hydrolysis of the alkylsiloxane moiety, treated with an O_3 solution (CH_2Cl_2) and washed with methanol to remove the PI component. Nuclear magnetic resonance (NMR) and infrared (IR) spectroscopic data confirmed both the crosslinking and degradation efficacies. The resultant films were opaque and fragile, and by gravimetric analysis, 70% of the PI was removed. Scanning electron microscopy (SEM) analysis on a cross section of the film revealed a layered-like pore structure with features on the order of 10 nm, consistent with some shrinkage of the film. Importantly, the authors reported the N_2 adsorption data for this porous film and concluded that indeed a continuous mesoporous structure was obtained ($74\ m^2\ g^{-1}$). A year later, in a more detailed full publication, Lee et al. reported similar results for a series of PPS-PI-PPS triblock copolymers with different morphologies (as evidenced by transmission electron microscopy [TEM]) [14]. The PI degradation reaction (ozonolysis) was not complete and the lengths of the residual PI chains in the porous materials were estimated based on gravimetric data. The authors suggested that these residual (and functionalized) chains coat the pore surfaces. Again, the porous structures were confirmed by SEM (Fig. 1). In the sample that contained spherical PI inclusions before degradation, the pore structure appeared to be "closed", consistent with a discontinuous structure. Interestingly, the ozonolyis-based degradation was quite effective at PI removal. This suggests that these domains were readily accessible to O_3 and that the degradation products could be extracted without difficulty. In addition to nitrogen adsorption (Brunauer–Emmett–Teller [BET]) analysis on these materials, the N_2 permeability data was provided for these porous membranes. Permeabilities consistent with the pore sizes in the membranes were recorded. Very low permeability was observed for the closed-cell film; values roughly three orders of magnitude smaller than for the corresponding lamellar films were measured.

This work was important for several reasons in addition to the original demonstration of the methodology for generating porous films from ordered block copolymer precursors. Demonstrating porosity necessitates more than just microscopy. The complementary N_2 adsorption, gravimetric, spectro-

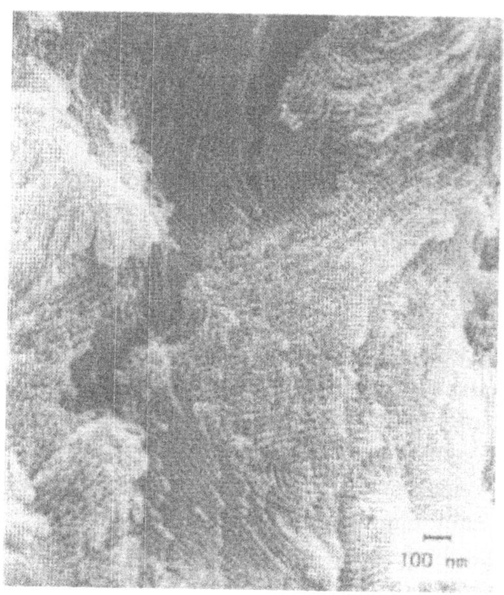

Fig. 1 Nanoporous membrane from a cylinder-forming PPS-PI-PPS triblock copolymer. Reproduced from [14]

scopic, and permeability data provided a compelling case for the proposed structures. Furthermore, the idea that the resulting pore structure could be dictated by the block copolymer precursor was not lost on these authors. In their 1988 publication they wrote: "Furthermore, it is observed that the resulting membrane contains carbonyl groups by oxidative cleavage of the polymer chain, which will enable some functions such as enzymes and metal complexes to be linked in the micropores."

In 1992 Smith and Meier published a paper on the degradation of the polydiene component in polystyrene-polybutadiene (PS-PBD) or PS-PI diblock copolymers [15]. This work was aimed at developing new techniques for the visualization of the three dimensional morphology of microphase separated block copolymers. While most of the work in this paper describes the degradative processing of polydiene-rich materials, one example of a symmetric (i.e., lamellar) material is given. Like the Lee et al. work described above, the degraded sample was probably porous, although no TEM image was shown for this material. The work reported by Smith and Meier demonstrated that the ozonolyis procedure can effectively remove the polydiene component without negatively affecting the uncrosslinked PS domains. While their aim was not to generate porous materials, the work is important for this reason.

Hedrick et al. elegantly demonstrated a block copolymer approach to generating porous materials, termed nanofoams, using polyimide-based materials in 1993 [16]. With the aim of preparing interlayer dielectrics with low dielectric constants to prevent "crosstalk" in microelectronic devices, polyimides were prepared that contained thermally labile blocks such as poly(methyl methacrylate) (PMMA) or poly(propylene oxide) (PPO). Upon heating these materials above the decomposition temperatures of the degradable blocks, the volatile byproducts are removed leaving behind voids in the inherently high glass transition temperature polyimide [poly(phenylquinoxaline) in this case]. The block copolymers synthesized in this (and subsequent work) were prepared by condensation polymerization techniques and gave microphase-separated structures that typically lacked a high degree of order. However, for the intended application, the key features of the precursor structures were that they needed to generate small, closed-cell voids. In this regard the materials generated by this approach were exemplary. Using a combination of characterization techniques, the targeted nanofoam structures were shown to be quite thermally stable and to have low dielectric constants and densities. In subsequent work, this group demonstrated the generality of the polyimide block copolymer approach for generating porous materials, which was recently summarized in a review entitled "Nanoscopically Engineered Polyimides", and the reader is referred to that publication (and references therein) for more detailed information [17].

A seminal contribution by Mansky et al. on the preparation of oriented thin films of PS-PBD block copolymers that were then subjected to ozonolysis degradation conditions to give porous thin films for use as nanolithography masks was published in 1996 [18]. While the ozonolysis was directly derived from the work of Lee et al. [13, 14] and Smith and Meier [15], the authors used a technique they developed to prepare thin films of block copolymers with cylindrical domains oriented perpendicular to the film surface. By drop casting toluene solutions of a PS-PBD block copolymer on the surface of deionized water, thin (ca. 100–200 nm) films were obtained. TEM analysis showed that upon annealing the films formed well-organized structures. Removal of the polybutadiene phase resulted in porous thin films as evidenced by TEM (Fig. 2). In this publication the authors also refer to successful experiments using these films as lithographic masks through pattern transfer techniques. This noteworthy work stimulated significant research in the area of "nanolithography" [19] using block copolymer templates as masks (vide infra).

In 1997 Liu et al. reported the synthesis, self-assembly, and hydrolysis of an interesting block copolymer [20]. Using an anionic polymerization methodology, poly(*t*-butylacrylate)-*b*-poly(2-cinnamoylethyl methacrylate) [PtBA-PCEMA] diblock copolymers were prepared, and the authors demonstrated formation of a microstructure containing PtBA cylinders hexagonally packed in a PCEMA matrix. Exposure of thin films (ca. 50 nm) of this material to UV irradiation crosslinked the PCEMA phase (35% of the double bonds were reacted). Immersing the films first in a CH_2Cl_2 solution of trimethylsilyl iodide followed by a methanol/water solution lead to cleavage of the *t*-butyl group leaving behind the corresponding poly(acrylic acid). Impressive TEM micrographs of thin films of these crosslinked/hydrolyzed films were obtained without any chemical staining (Fig. 3). The cylindrical features were light in the transmission image suggesting that the density of the cores was less than that of the crosslinked PCEMA matrix. In principle, the material generated was at least semi-porous, although no other evidence was reported in this manuscript. Nonetheless, this work showed that partial removal of the minor-

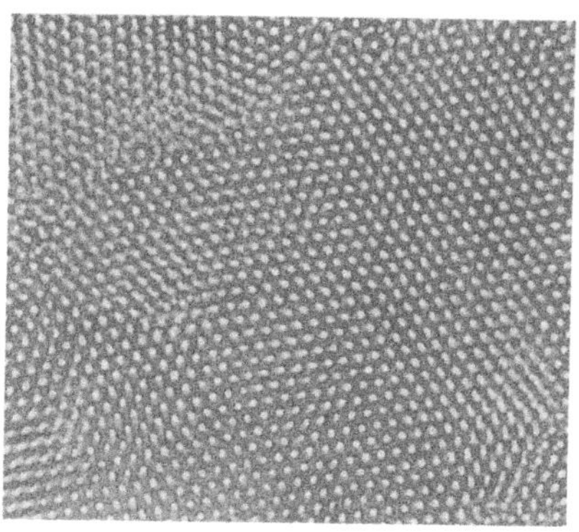

Fig. 2 TEM image (without staining) of a nanoporous PS membrane formed by ozonolysis and washing. Reproduced from [18]

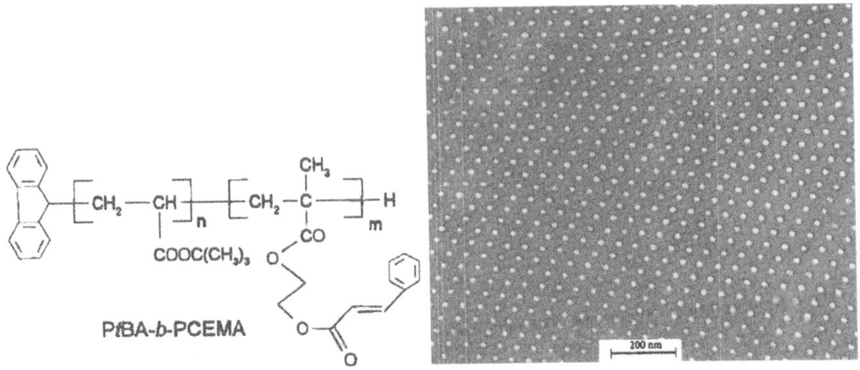

Fig. 3 Structure of PtBA-PCEMA and TEM image after hydrolysis. Reproduced from [20]

ity component of a block copolymer by a simple hydrolysis could be achieved, and therefore could be used as a straightforward protocol for the generation of porous materials. Subsequent work by this group demonstrated that this was indeed the case (vide infra).

Hashimoto et al. reported the ozonolysis/degradation of a PS-PI block copolymer/PS homopolymer blend to generate nanoporous films in 1997 [21]. The key feature of this work that differentiates it from related prior schemes is the morphology of the block copolymer precursor. In this case, the film adopted a bicontinuous morphology with $Ia\bar{3}d$ symmetry, or the so-called gyroid phase. This morphology consists of percolating networks of both PS and PI domains. Because of the inherent bicontinuity, orientation of the morphology is not required if porous versions of this phase are to be used for membrane-based separations. Relatively thick (100–300 μm) films of the blend were cast by slow evaporation from toluene. The gyroid morphology was confirmed by TEM. Exposure of the films to ozone followed by treatment with ethanol (a nonsolvent for PS) resulted in complete removal of the PI phase by NMR spectroscopy and gravimetric analysis (although the details were not reported in this paper). Complete removal of the PI phase in such a thick film using ozonolysis is significant, and further supports the formation of a three-dimensionally continuous material. SEM images of the porous materials were consistent with the proposed structure (Fig. 4). In fact, the symmetry of the precursor material was nearly identical to the resultant porous material. The "nanochannels" could be coated with nickel by a non-electrolytic plating technique. In the Ni-coated materials, metal nanoparticles were deposited in the channels, and the authors propose that these particles form essentially a continuous coating. The authors make the comment that continuous channels coated with catalytically active metals could prove to be quite useful as membrane reactors.

These contributions between 1988 and 1997 set the stage for a wide variety of contributions in the field of porous materials derived from block copoly-

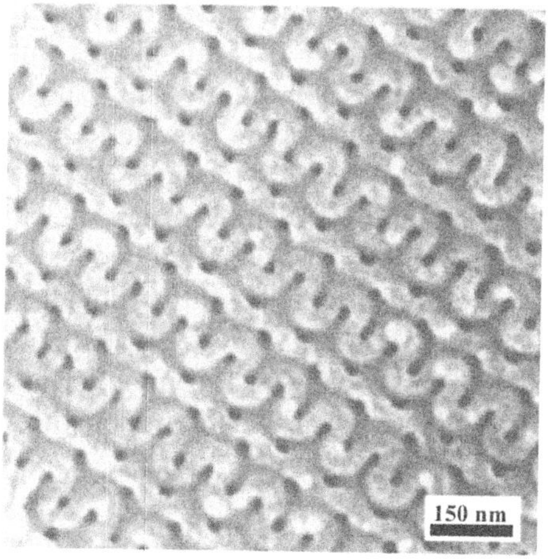

Fig. 4 Nanoporous membrane from an etched gyroid structure. Reproduced from [21]

mer precursors. Much of the work described in the following sections can be traced back to the visionary work of the Nakahama group in 1988 [13]. The research reported in that 1988 publication evolved from the development of anionic polymerization methodologies that allowed for the incorporation of functional groups (e.g., silyl groups). The design and development of new synthetic methodologies often spurs other tangential discoveries that find utility in a variety of technological applications. The generation of porous materials from "functionalized" block copolymer precursors is evidence of this. Porous materials from ordered block copolymer precursors have been developed for various applications since 1988. From interlayer dielectrics to nanostructure templates to selective membranes these materials have received considerable attention over the last decade.

3
Block Copolymer Lithography Employing Nanoporous Templates

In what has become one of the most cited papers in the field of block copolymer lithography, Park et al. described tactics for preparing either pits or posts of SiN using block copolymer masks and a combination of chemical modification and reactive ion etching (RIE) [22]. In the protocol for preparing holes (pits) in a SiN substrate, a PS-PI or PS-PB block copolymer was spin coated on a SiN substrate and subsequently annealed. Monolayers of these materials formed spherical inclusions of the polydiene component. Using the established O_3 degradation technique, the authors show that the polydiene phase can be etched out leaving behind a mask that has periodic material density contrast as a result of the porous structure (Fig. 5). Subsequent RIE using a fluorocarbon-based plasma results in anisotropic etching of the substrate. In one example, SEM analysis (Fig. 5) of the etched film (after removal of a surface coating of PS) and the resultant pitted SiN substrate clearly demonstrates the efficacy of this double etching technique (i.e., a chemical etch of the diene, followed by RIE). In this case the holes were approximately 20 nm in diameter with a period of about 40 nm. The natural length scale that block copolymers self-assemble on leads to remarkably dense arrays, and this fact was highlighted by the authors in the title of their manuscript "Block Copolymer Lithography: Periodic Arrays of $\sim 10^{11}$ Holes in 1 Square Centimeter". The authors note the advantages of block copolymer lithography over electron beam lithography in this work and demonstrate the benefit of being able to pattern large areas.

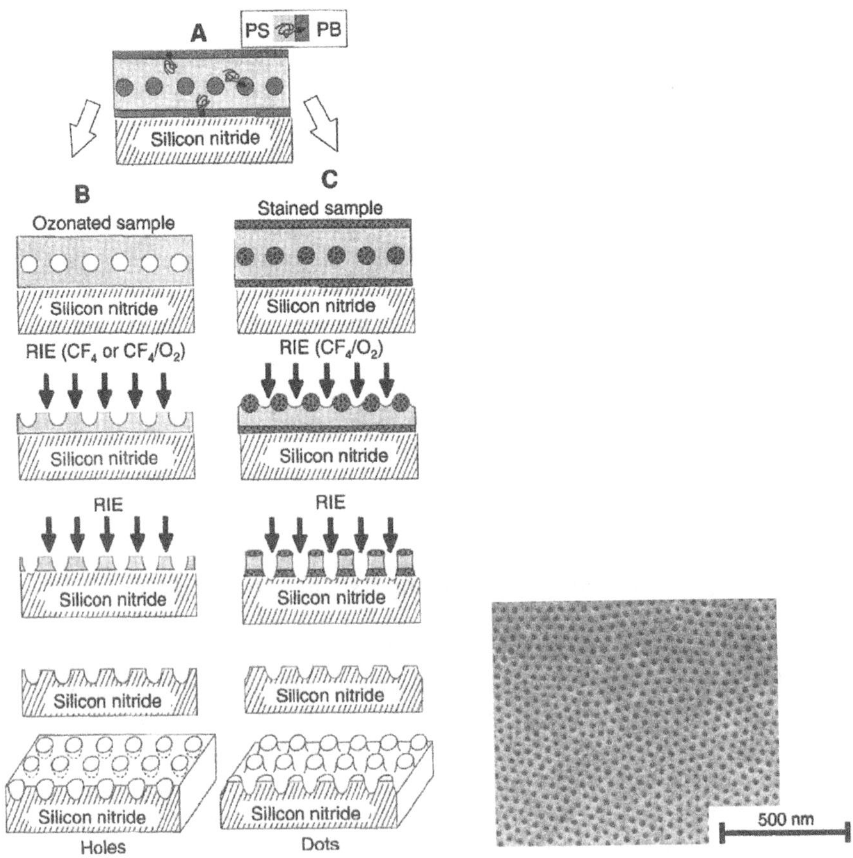

Fig. 5 Protocol for preparing nanoporous thin films for nanolithographic applications (route B) and SEM image of the resultant nanoporous thin film (after removal of the PS overlayer). Reproduced from [22]

In a follow up full publication to the work above, Harrison et al. showed that using porous thin films as described above, pattern transfer could be accomplished in both Si and Ge substrates as well [23]. While an entire three-inch Si wafer was patterned using this technique, practical limitations, such as in etch anisotropy, were reported. In this paper, an effective technique for confirming porosity in these films was demonstrated. After ozonolysis of these thin films, an overlayer of PS remained at the surface of these porous samples. By using a low power CF_4 RIE followed by periodic SEM analysis, this layer could be slowly etched to reveal thus revealing the cylindrical pores (now observed as trenches). Also, the authors showed that the low power CF_4 etch did not result in significant surface roughness.

Limitations of the porous polymer mask approach described above, when attempting to use it for metal deposition (rather than simple pattern transfer to the underlying substrate), include the small aspect ratio of the holes and mask lift off problems. To circumvent these issues, Park et al. developed a tri-layer pattern-transfer method using a polyimide, SiN and block copolymer layers on arbitrary surfaces (Fig. 6) [24]. After ozonolysis of a block copolymer thin film containing spheres of PI in a matrix of PS, the pattern was transferred into the SiN layer using the CF_4 RIE protocol (as in this group's previous work). In the trilayer scenario, the patterned SiN can be transferred to the polyimide layer using an O_2 RIE. The etch anisotropy between the inorganic SiN and the organic polyimide is quite large. Essentially, the SiN is an excellent mask for O_2 etching and deep holes can be generated without compromising the pattern. While there was some lateral widening of the holes at aspect ratios greater than about three, this trilayer technology was quite ef-

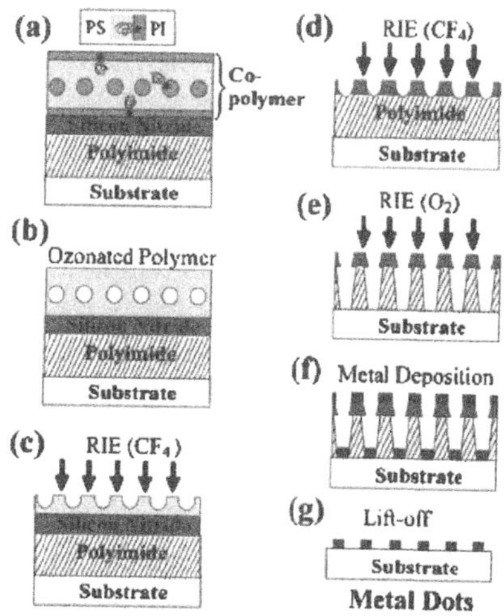

Fig. 6 Trilayer pattern transfer approach using nanoporous diblock copolymers. Reproduced from [24]

fective for preparing, for example, Ti/Au nanodot arrays. The authors go on to propose the generic utility of this protocol for nonplanar substrates, electroplating techniques, and micro(nano) molding. In the concluding remarks the following comment was noted: "We expect that slight modifications of this technique will provide an inexpensive, fast, and versatile means of producing dense periodic structures at the 3–100 nm scale over large areas on any substrate of interest".

In the work with thin films of PS-polydiene block copolymers described above, the morphology of the materials was exclusively spherical in the thin film geometries. Also, a top layer of the low surface energy block was always present and the spherical inclusions were buried in the film. Thus the masks generated behaved in an anisotropic manner under etching conditions. An ideal mask, however, would have channels that were open on the surface of the film and extended through the entire film to the substrate. Porous materials of this nature would allow for direct deposition (or growth) of material in the holes, and following lift off, leave the corresponding patterned surfaces. One way to achieve this porous mask using block copolymers is to employ cylinder-forming materials that could be manipulated in such a way as to allow for the cylinder orientation perpendicular to the substrate surface. Subsequent removal of the component that formed the cylinders would leave the nanolithographic masks. In 2000, Thurn-Albrecht et al. demonstrated that this was possible using PS-PMMA block copolymers (prepared by anionic polymerization) [25]. Spinning a thin (40 nm) film of a PS-PMMA block copolymer containing 30 vol% PMMA onto a substrate coated with a PS/PMMA random copolymer gave a beautifully oriented thin film of PMMA cylinders standing on end and hexagonally packed in a PS matrix after annealing. The random copolymer film was used to render the surface ambivalent toward adsorption of either the PS or PMMA components in the block copolymer; this resulted in a thermodynamic preference for the perpendicular orientation [26]. Thicker films with this same orientation were generated using electric field alignment techniques developed by the same group [27]. Exposure of either of these films to UV radiation led to both degradation of the PMMA phase and crosslinking of the PS phase. Subsequent washing with acetic acid gave porous templates that exhibited the size, shape, orientation, spacing, and packing of the parent materials as confirmed by SEM (Fig. 7), TEM, atomic force microscopy (AFM), and small-angle X-ray and neutron scattering (SAXS and SANS). A key component of this work was the simultaneous crosslinking and degradation of the PS and PMMA, respectively. For thin films, a simple five-step procedure consisting of substrate coating (i.e., with random copolymer), block copolymer deposition, thermal annealing, exposure to UV light, and film washing/drying gives porous masks that are useful for subsequent nanofabrication techniques [28].

Eight months after submission of the first manuscript describing successful preparation of nanolithography masks from PS-PMMA block copolymers,

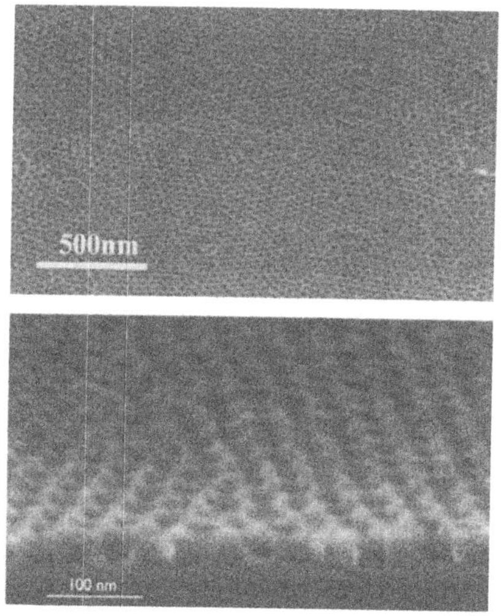

Fig. 7 SEM images of nanoporous thin films from cylinder-forming PS-PMMA diblock copolymers. Reproduced from [25]

Thurn-Albrecht et al. submitted a second communication describing the successful use of these porous templates for the preparation of "ultrahigh-density" arrays of cobalt nanowires [29]. These nanowires were grown in the pores (14 nm in diameter) of a one-micron thick template generated by electric field alignment of a PS-PMMA block copolymer followed by UV exposure, rinsing and drying. The wires were grown to a length of about 500 nm and left in the porous (and crosslinked) PS matrix (Fig. 8). The cobalt nanowires showed enhanced coercivities as compared to continuous cobalt films [30], providing promise for use of these arrays in ultrahigh-density storage (approaching one terabit per square inch). These two publications are noteworthy. While both the methodologies employing sphere forming block copolymers and cylinder forming block copolymers each have their respective merits, ideal templates for nanolithographic applications are the type described using the PS-PMMA materials. While one must turn to electric field alignment to prepare oriented films of arbitrary thicknesses, the resultant porous templates are truly exemplary masks.

After these initial reports, this group has published no less than a dozen other papers utilizing and expanding on this versatile technology that exploits porous materials derived from block copolymer precursors. As suggested in their *Science* report [29], thin porous templates prepared from a PS-PMMA material were used by Kim et al. as "nanoscopic reaction vessels" for the growth of silicon dioxide pillars on a silicon wafer [31]. SiO_2 was grown (by hydrolysis of $SiCl_4$) in the pores generated by UV exposure (and acetic acid rinsing) of a thin PS-PMMA film that was cast on a random copolymer substrate and annealed. After a CF_4 reactive ion etch the

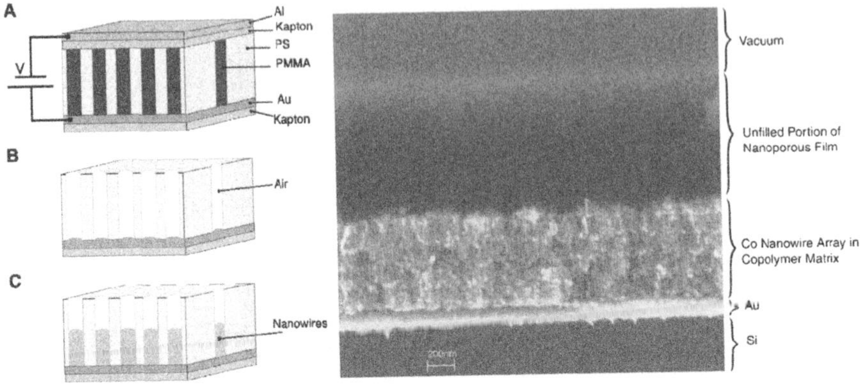

Fig. 8 Schematic of nanofabrication methodology (*left*) and SEM image of Co nanowire array (*right*). Reproduced from [29]

remaining polymer was removed along with some of the SiO_2, although the etching rates were different enough to provide sufficient contrast. SEM image showed the remaining hexagonally packed SiO_2 posts 6 nm high and 20 nm in diameter with a lattice constant close to that of the precursor material. Although the authors do not state this explicitly, these results suggest that in fact the random copolymer layer used to induce perpendicular orientation of the cylinders in the PS-PMMA template is removed (i.e., degraded) during the UV/acetic acid treatment. Thus, bare SiO_2 is exposed at the bottom of the pores. For templating applications this is an important aspect of the overall PS-PMMA mask procedure.

In related work, Black et al. showed that these porous PS masks could be used to "drill" pits into a silicon wafer through exposure to an SF_6 plasma [32]. With an estimated etch selectivity of 25 : 1 (Si : PS), the mask was quite effective for this process. Integration of this procedure with standard microelectronic processing techniques was used to generate metal-oxide-semiconductor capacitors with increased charge storage capacities, important for dynamic random access memory applications. Guarini and coworkers followed up on this initial report with two studies describing in more detail the use of porous PS templates as nanoscale patterning media for semiconductor applications. In 2001 Guarini et al. showed that ordered porous templates could be formed across entire 8-inch diameter silicon wafers [33]. Some optimization of the block copolymer spin casting procedure and subsequent annealing protocols was disclosed. In this work, the random copolymer coating approach was used to obtain the preferential orientation. PS-PMMA films between about 33 and 42 nm thick, annealed at 160 °C for about a day and a half gave the best order. As described in previous work by this group, SF_6 RIE and metal deposition methodologies were reported. An optimized procedure for the deposition of gold nanodot arrays using the PS masks was also

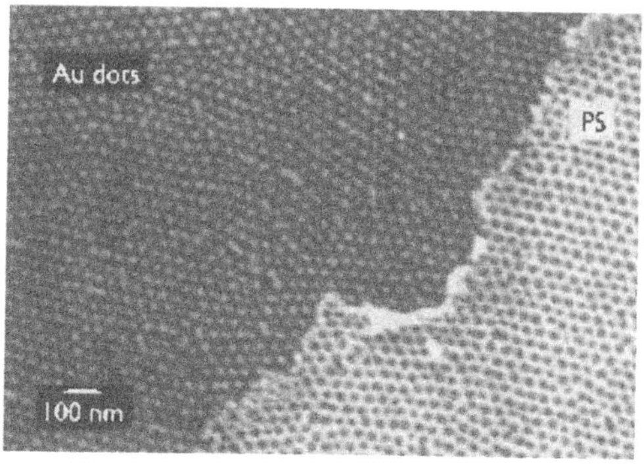

Fig. 9 Gold nanoparticles (*left hand side of the image*) prepared by deposition onto a nanoporous PS mask derived form a PS-PMMA block copolymer. Reproduced from refernence [33]

beautifully illustrated (Fig. 9). These arrays were generated by simply evaporating gold onto the mask followed by lift off (sonication in organic solvent). Using a virtually identical process, Shin et al. reported the preparation of gold, chromium, and multilayer (Cr/Au) nanodot arrays in 2002 [34]. Notably, nanoporous metal films were prepared using a block copolymer with the majority of the PMMA phase thus generating an array of PS pillars upon UV exposure and acetic acid rinsing. In 2002 Guarini et al. showed that the nanoscale patterns provided by the porous PS templates could readily be transferred into "more robust" materials for use in semiconductor nanofabrication processes using RIE-based and wet etch processes (with various gases) [35]. This work and subsequent reports by Jeong et al. in 2001 [36] and Bal et al. in 2002 [37] further demonstrated the utility and versatility of these PS masks for the potential and realized fabrication of nonvolatile memory devices, lasers, biochips, field-emission displays, nanoelectrode arrays [36], and magnetotransport devices [37].

The effectiveness of nanoporous membranes prepared from ordered PS-PMMA block copolymers certainly motivated in-depth work on optimizing the fidelity of the structures and controlling the feature sizes. In two papers describing detailed and careful characterization of PS-PMMA thin films, the self-assembly characteristics were well documented. In 2001 Xu et al. demonstrated that using the balanced interfacial energy approach to perpendicular cylinder orientation (i.e., using random copolymer), PS-PMMA materials with molecular weights up to 103 kg mol^{-1} gave nicely ordered films upon annealing [38]. However, a 295 kg mol^{-1} PS-PMMA sample showed poor order under the same conditions. This was attributed to the slow collective motion of the polymer chains required to achieve a well-ordered sample. Using an optimized electric field alignment technique, all PS-PMMA films studied gave perpendicular microdomain orientation. The authors suggest that the forces present as a result of the external field overwhelm kinetic limitations in even the highest molecular weight PS-PMMA sample examined. An experimental scaling of domain spacing (D) with degree of polymerization (N) was consistent with strongly segregated PS-PMMA above about 80 kg mol^{-1} ($D \sim N^{2/3}$). Thus the range of pore sizes that could be generated is from 14 to 50 nm.

Guarini et al. examined the influence of film thickness, annealing time, annealing temperature, and block copolymer molecular weight on the porous structures resulting from degradation of PS-PMMA thin films spun on a PS/PMMA random copolymer coated substrate ("to neutralize the substrate") [39]. From this work, there appears to be an optimal film thickness centered at about 38 nm and an optimized annealing time and temperature (1 h, 180 °C) for a 67 kg mol^{-1} PS-PMMA material containing 70 wt% PS. SEM images of the porous films were analyzed in a quantitative manner using image analysis software. With a block copolymer of higher molecular weight (132 kg mol^{-1}), these authors reported a scaling of the domain spacing with N of 0.58, in general agreement with Xu et al. [38]. These authors also reported that the higher molecular weight PS-PMMA materials gave more poorly ordered porous films (Fig. 10). Diminished diffusivity of the high molecular weight sample as compared to the 67 kg mol^{-1} sample was implicated.

As demonstrated above, the use of PS-PMMA block copolymers as precursors to nanoporous materials is quite well established. Electric field alignment

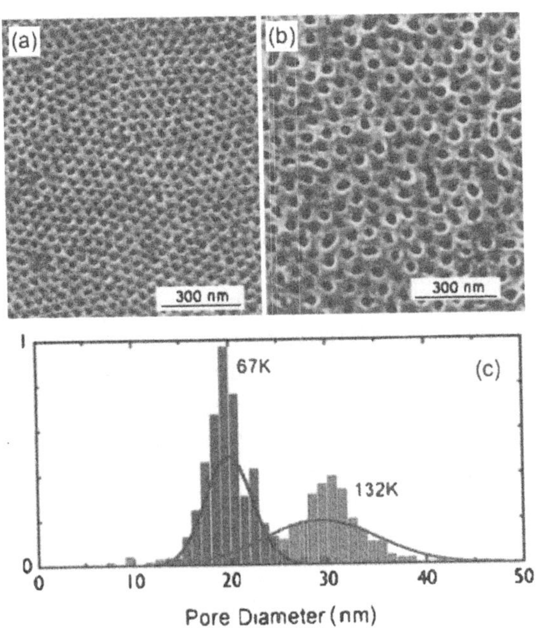

Fig. 10 SEM of nanoporous PS materials showing the influence of molecular weight on the ordering of templates **a** 67 kg mol^{-1} **b** 132 kg mol^{-1} **c** pore diameter histograms. Reproduced from [39]

or surface modification gives cylindrical features that orient perpendicular to the substrate surface. Crosslinking the PS and degrading the PMMA gives the porous materials. Several variations on this theme have been reported that utilize other strategies to generate the final porous materials from the same precursor blocks. In 2002, Jeong et al. described the generation of nanostructured PS-PMMA (83 kg mol^{-1}) films that contained up to 10 wt % PMMA homopolymer (18 kg mol^{-1}) [40]. These authors suggested that the PMMA homopolymer preferentially segregates to the center of the cylindrical PMMA nanodomains. Therefore, simple dissolution of the PMMA homopolymer in a nonsolvent for the PS (acetic acid/water mixture) leads to the generation of voids (6 nm in diameter) in the film. By examining a film that was folded back onto itself, the authors claim that the pores span the entire film (i.e., from surface to substrate). Addition of homopolymer to the PS-PMMA films allows for a quite simple technique to generate nanoporosity and controlling pore size. In particular this methodology can be used to generate pores that are quite a bit smaller than can be easily prepared in the pure PS-PMMA materials.

In a later report in 2003, Joeng et al. mentioned that while the homopolymer strategy was effective, the resulting nanoporous materials were not stable to elevated temperatures [41]. Since the matrix PS phase is not crosslinked (or otherwise modified) upon simple exposure to a nonsolvent, the PS sustains the pore structure due to its glassy (and high modulus) nature. At high temperature, the porous structure is compromised. In this publication [41], Jeong et al. show, that this limitation can be overcome by first crosslinking the PS by exposure to ozone at 17 °C [42]. The PMMA homopolymer (up to 30 wt % in this case) could then be removed by the aforementioned dissolution technique (in this case pure acetic acid is used). The films could then be annealed at 170 °C (well above the glass transition temperatures of both PMMA and PS) and retain the porosity as templated by the PMMA homopolymer. A lovely series of TEM images demonstrated the control of pore size and stability of the porous structure to high temperatures. Geometrical arguments are given in this work and lend theoretical support to the pore sizes observed. By combining this ozone crosslinking/homopolymer PMMA removal method with the standard UV degradation methods, these authors were able to generate pores ranging from 6–32 nm and predict that with homopolymer molecular weight variation this range can be significantly extended.

The ozone-induced crosslinking of PS in a PS-PMMA thin film in and of itself has produced nanoporous materials as demonstrated by Jeong et al. [42] in 2003 using a crosslinking/volume reduction approach. Ozone treatment under the proper conditions leads to crosslinking of PS and no significant modification of the PMMA. Of course most crosslinking chemistry leads to an increase in the density of the phase affected. In the case of ozone-induced crosslinking of PS, the relatively minor (although undetermined) mass increase is not significant, and the reduction in the volume dominates the change in density. Simple exposure of an oriented PS-PMMA thin film to ozone for a controlled time at 17 °C followed by heating to 170 °C generates porosity in the film. Increasing the ozone exposure time leads to an increase in the pore size (from 3 to 8 nm). The authors suggested that in a PS-PMMA thin film in which the PS phase is crosslinked, reduction of the film volume occurs only in the direction perpendicular to the substrate surface (i.e., the film thins without visible lateral contraction). When the films are heated above the glass transition temperature of the PMMA phase (not crosslinked, but made "uncomfortable" by the stress being applied by the surrounding crosslinked PS domains), the authors argue that the PMMA chains can relieve this stress by flowing toward the edges of the cylinder walls. Thus in the PS-PMMA block copolymers, the PS phase crosslinking results in reduction in volume of the film that is not laterally constrained on a local length scale. That is, the PS phase in the PS-PMMA phase becomes thinner and contracts in the lateral direction by pulling on the PMMA chains essentially voiding that material (not unrelated to rubber cavitation in toughened plastics during fracture). In fact, the ozone treatment leads to no contraction in PMMA volume in homopolymer films, and in PS homopolymer films the volume contraction can be used to predict the size of the voids in the crosslinked and annealed PS-PMMA samples. The arguments presented by the authors are confirmed experimentally through this correlation. This is quite a clever method for generation of porosity in thin films.

Of course reducing the number of chemical (e.g., UV, ozone, and/or solvent exposure) and thermal processing steps needed to generate nanoporosity in block copolymer thin films is desirable, and the last two techniques described with PS-PMMA films represent progress towards that goal. In the Fall of 2003 Xu et al. reported a very simple technique for generating nanoporosity in thin films of PS-PMMA [43]. Exposure of perpendicularly oriented thin films to acetic acid/water mixtures without any prior UV exposure (as was done in the first reported examples [25]), leads to nanoporosity as determined by AFM. The proposed mechanism for pore formation is that the PMMA chains are "drawn" to the film surface through selective dissolution in the solvent. Solvent treatment was shown to have no detrimental impact on the molecular weight of the chemical make up of the PS-PMMA materials (by GPC and NMR spectroscopy). The pore generation is reversible (and can be cycled); heating the film regenerates the dense film and the topographic AFM image of the film shows no features. A key issue for this process is the depth of the pores generated. Substrate TEM images could be obtained floating the thin films off of the substrate. In the TEM image presented, cylindrical bright areas were observed and the authors state: "The bright spots were the same intensity as if no sample were present in the beam ...". Thus the conclusion that the "pores extend completely through the film" was made. However, the authors do not comment on the layer of end-functionalized random PS/PMMA material that was coated on the SiO_2 before deposition of the block copolymer. From what was reported, the film they examined by

TEM was coated with 6 nm of the random copolymer and 34 nm of the block copolymer. Annealing at 170 °C and exposure to acetic acid generates the desired nanoporosity. After floatation of the film on a dilute aqueous solution of HF the authors picked it up on a TEM grid. The fate of the random copolymer was not described. Water contact angle and X-ray photoelectron spectroscopy (XPS) experiments were consistent with the migration of the PMMA chains to the top of the thin film to generate the desired porosity.

Increasing the long-range lateral order in thin films of block copolymers is important for applications that require some form of addressability (e.g., magnetic hard drives). Long-range order in a block copolymer thin film that was designed as a nanoporous polymer precursor could ultimately result in nanoporous materials with identical long-range ordering, and therefore this has been a goal in the field. While there are several methods for precisely defining the position of a nanodomain in a block copolymer thin film [7], in 2004 Kim et al. reported that simple solvent annealing of PS-poly(ethylene oxide) [PEO] thin films lead to remarkable arrays of highly ordered PEO cylindrical domains in a PS matrix over quite large areas ($5 \times 5\,\mu m^2$) as shown in Fig. 11 [44]. In fact, only one defect was observed in the $25\,\mu m^2$ AFM image. The mechanism for generating these arrays is presented as being due to a concentration gradient established in the thin film upon solvent evaporation post annealing. Later that same year, Kim et al. showed that this solvent annealing process could be applied to thin films of diblock/homopolymer blends (PS-PEO/PEO and PS-PEO/PMMA) to

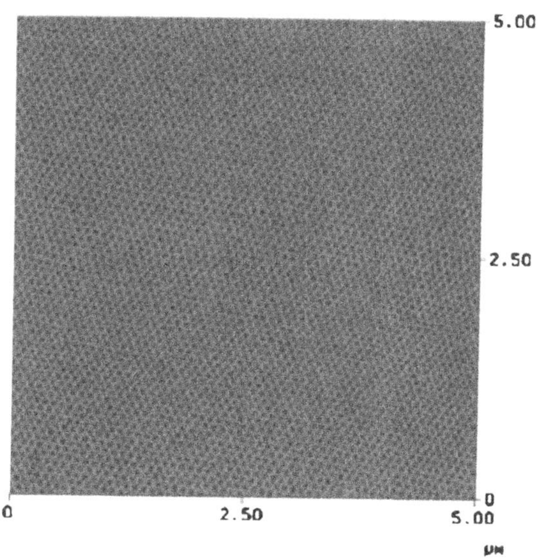

Fig. 11 AFM image of a 260 nm thick film of PS-PEO cast from benzene in a benzene/water atmosphere for 48 h. Reproduced from [44]

give similar results, and went on to demonstrate by both TEM and grazing-angle small-angle X-ray scattering (GASAXS) that the cylindrical domains span the entire thickness of the film [45]. Importantly, these authors showed that nanoporous materials could be generated from these highly ordered arrays by simple homopolymer removal (i.e., dissolution of PMMA in a PS-PEO/PMMA blend) as described above [40]. This combination of solvent annealing and selective dissolution elegantly shows that by simple manipulation of block copolymer thin films, nanoporous materials can be generated with narrow pore size distributions and remarkable long-range order.

The solvent exposure techniques described above, the UV degradation of PMMA, and the removal of PMMA homopolymer by simple dissolution all require the use of solvents (wet etching techniques). In 2002, Asakawa et al. demonstrated that dry-etching techniques could be used to selectively remove the PMMA phase from PS-PMMA block copolymers, leaving pores in the material that could be used to generate nanoscopic features in the underlying substrate [46]. Using PS-PMMA block copolymer/PS homopolymer blends, these authors generated thin films of PMMA spheres in a PS matrix on hydrophobically modified silicon wafers. A rather detailed account of preparing these films with different molecular weights and compositions was presented. Given the disparate resistance of these two materials to oxygen plasma reactive ion etching (O_2-RIE), nanoporous films could be generated by the selective removal of the PMMA phase. This was demonstrated by the following three step process: O_2-RIE (selective removal of the PMMA), CF_4-RIE (selective etching of the underlying Si), and extended O_2-RIE (complete removal of the entire polymer). The resulting pitted Si surfaces provided evidence for a nanoporous PS intermediate. The authors tout this three-step process ("coat, anneal, and etch") as a very simple method for generating nanoscale patterns. In fact, the same year Naito et al. used a related procedure to manufacture a 2.5-inch disk patterned with magnetic nanodots [47, 48]. In that work a PS-PMMA block copolymer was spin cast onto a substrate with circumferentially patterned grooves. Removal of the PMMA by O_2-RIE etching was followed by filling with a "spin-on-glass" (SOG), a solution consisting of predominantly methyl siloxanes (OCD Type-7, Toyko Ohka) [49]. The spin-on-glass filled holes were reported to act as masks for subsequent ion beam milling. The resulting structures were well-aligned arrays of magnetic nanodots (40 nm in diameter) in the pre-patterned groves of the disk. Magnetic characterization of the dots by conventional magnetometry and magnetic force microscopy after both ac and dc erases confirmed single magnetic domains. The technology described in this publication was described by Park et al. as "likely one of the first commercial applications of block copolymers in nanotechnology" [7].

In 2004, Olayo-Valles et al. described a related dry-etch processing of self-assembled block copolymer films to generate porous materials for use as magnetic material templates [50]. These authors employed thin films of

PS-polylactide (PLA) block copolymers spin coated on a variety of substrates (SiO_2, Al_2O_3(0001), MgO(100), GaAs(001), and Cu) that formed PLA cylinders in a matrix of PS. Upon thermal annealing the PLA cylinders in these thin films oriented perpendicular to the substrate surface in all cases. This orientation was argued to be the result of no preferential wetting by either of the components at the air-polymer interface due to the similarity in the surface energies of these two polymers. By simple exposure to aqueous base, the PLA phase was etched leaving pits in the film (confirmed by SEM and AFM). However, the aqueous treatment led to delamination of the thin film. To circumvent this problem, these authors stained the films with RuO_4 vapors, a selective stain for the PS phase. Upon exposure of these films to an O_2-RIE the Ru-stained PS etched very slowly (the putative ruthenium oxides formed presumably act as an etch stop) relative to the PLA cylinders (in this example unstained PS and PLA had nearly the same etch rates). Porous thin films resulted, and molecular beam deposition of metal onto these masks followed by template lift off gave nanodot arrays of, for example, $Ni_{80}Fe_{20}$ (Fig. 12). This magnetic nanodot array exhibited enhanced coercivity and reduced squareness as determined by magnetometry (isothermal hysteresis loops). Since this processing can be done on crystallographically oriented substrates, the generation and study of epitaxial nanodot arrays was envisioned.

Two papers appeared in 2004 describing an interesting combination of "phase-selective" crosslinking chemistry and polymer degradation to generate nanoporous thin films from patternable block copolymer systems. Du et al. demonstrated the generality of this approach using three different (but related) block copolymer systems [51]. One of these, a poly(α-methyl styrene)-poly(4-hydroxy styrene) [PαMS-PHOST] system, was described in more detail by Li et al. [52]. The overall idea in this approach is to obtain spatially controlled nanopores in thin films using combined lithographic and

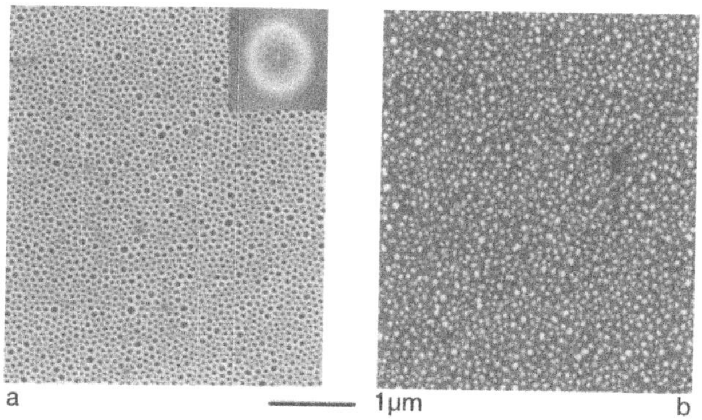

Fig. 12 SEM image of **a** a RuO_4 stained PS-PLA film and **b** a $Ni_{80}Fe_{20}$ nanodot array formed from the resultant nanoporous PS template. Reproduced from [50]

polymer degradation methodologies (Fig. 13). Thin films of PαMS-PHOST (prepared by anionic polymerization and postpolymerization modification) containing a PHOST-selective crosslinking agent (tetramethoxymethylglycouril) and a photoacid generator (triphenylsulfonium trifluorosulfonate) were prepared by spin coating. Cylindrical domains of PαMS oriented perpendicular to the substrate surface were observed. Because of the high T_g of the PHOST and the low ceiling temperature of the PαMS, UV irradiation of these films (to generate free radicals) at 80 °C led to depolymerization of the PαMS. By combining the lithographic characteristics of the PHOST block with this degradation of the PαMS, the authors were able to show that micron-sized patterns of the PHOST-PαMS block copolymer could be generated by a standard lithographic process (negative-tone photoresist technology). In a second step, the block copolymer patterns could be rendered nanoporous by depolymerization of the PαMS phase. AFM images showing the precise control enabled by this approach are shown in Fig. 14. This marriage of lithography and nanoporosity generation is quite powerful and will likely find use in various applications as described by the authors.

A unique methodology for the generation of nanoporous thin films using supercritical carbon dioxide ($scCO_2$) for potential nanolithographic applications was introduced in 2004 by Li et al. [53]. In this work, thin films of PS-poly(perfluorooctylethyl methacrylate) [PFMA] were spin-cast onto sili-

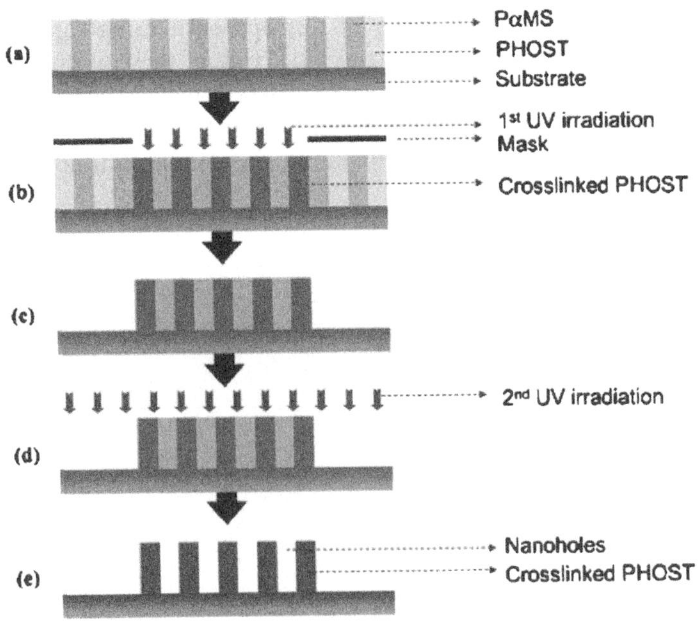

Fig. 13 Nanofabrication process for obtaining spatially controlled nanopores. Reproduced from [52]

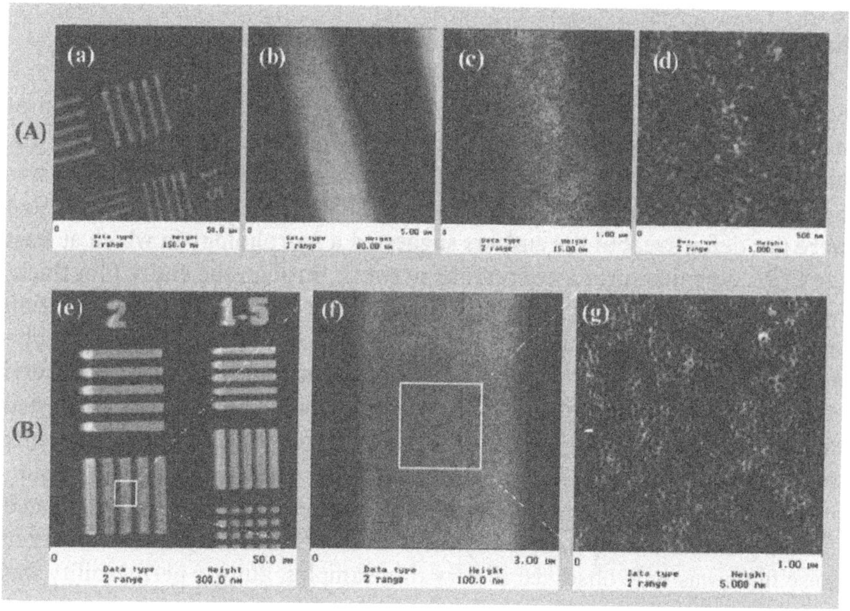

Fig. 14 AFM images showing photopatterned lines of nanoporous materials. **A** is for a film thickness of 39 nm and **B** is for a film thickness of 127 nm. Reproduced from [52]

con wafers and shown to contain PFMA "nanodomains" (likely a single layer of spherical inclusions of PFMA in a continuous PS matrix). The films were then treated with $scCO_2$ at elevated temperature resulting in selective swelling of the CO_2-philic PFMA domains and an overall increase in film thickness. After an appropriate temperature quench, the PS matrix vitrified and subsequent depressurization led to empty nanocells "coated" with the low surface energy PFMA. After a RIE step, "pitted" thin films were obtained with 10^{10} cells per cm^2 (diameters of approximately 30 nm). These structures can, in principle, be used as nanolithographic masks, low-dielectric films, or nanoreactors. The authors also described how to control the size of the nanocells by simply controlling the $scCO_2$ pressure, allowing for a relatively simple tuning protocol. This process for generating nanoporosity in block copolymer thin films is quite unlike any other since the voids are templated by both a discrete phase segregated material and an added, selective "solvent". The $scCO_2$ not only allows for expansion of the film, but also acts as a porogen of sorts. This methodology was also expanded to monolithic nanoporous materials as described below.

4 Nanoporous Membranes for Separation and Photonic Applications

Following on their work in 1997 [20], Liu et al. reported the formation of nanoporous membranes from ABC triblock copolymers of PI-PCEMA-PtBA in 1999 [54]. A blend of this triblock copolymer and PtBA homopolymer was prepared and the morphology analyzed by TEM. The TEM images generated were not definitive, but the authors suggested a morphology in which at least the PtBA domains form a continuous network through the entire film thickness (in the microtomed samples). Crosslinking the film by UV irradiation lead to materials that could be "voided" by removal of the PtBA homopolymer (also employed by Jeong et al. in [40]), and/or hydrolysis of the *tert*-butyl esters. After extraction of the PtBA homopolymer, TEM analysis suggested that pores were generated in the material. A tortuosity of the pore structure was suggested. Films in which the homopolymer PtBA was removed by simple dissolution or by both PtBA removal and ester hydrolysis were evaluated in both gas and water permeability measurements. The water permeability of these films was only observed in very thin samples, but gas permeability was documented in thicker samples for a variety of gases. The main conclusion from these measurements was that the films were remarkably more permeable than, for example, polyethylene (6 orders of magnitude increases) and that gas molecules permeated the membranes according to a Knudsen flow model as evidenced by a linear relationship between the permeability of the films and the square root of the gas molecular mass. While detailed morphological analysis of this system was not presented in this manuscript, the data reported importantly established the efficacy of this type of membrane for potential separation applications.

In 1999 Liu et al. published a paper in which they showed that porous membranes generated by the same methodology they reported in 1997 [20] could be used as selective permeation membranes and "nanoreactors" for the formation of inorganic nanoparticles [55]. The PtBA-PCEMA material used in this publication formed nanoscopic cylinders of PtBA in a matrix of PCEMA. Crosslinking the PCEMA and hydrolysis of the PtBA gave nanoporous membranes (definitive SAXS and TEM analysis were presented). In the dry state actual voids were observed by TEM; when exposed to water the voids contained swollen polyacrylic acid chains. Thorough characterization of these membranes was presented. Interestingly, the permeability of these porous membranes showed three distinct regions depending on pH. At both high (13) and low (1) pH the membranes exhibited high permeability, but at intermediate pH values (3), the membranes showed quite low permeability (several orders of magnitude lower). The authors present arguments for these observations based on gel formation, high chain grafting density and water solvation. Also, the impact of ionic strength and the influence of divalent cations on the permeability were demonstrated. In fact, a 1 M solution of calcium chloride did not pass through the membrane. Polyacrylic acid crosslinking was implicated. On the basis of these results, the authors suggest that these membranes could be used as sensors or "chemical valves". Some of this data was also presented in a related publication a year earlier [56]. In that publication, Liu et al. made a brief comment about the size selectivity of related membranes by the exclusion of crosslinked PS-PCEMA micelles.

In 2001, the aforementioned PCEMA crosslinking/PtBA hydrolysis methodology was expanded to include the preparation of nanoporous microspheres. Using an amphiphilic block copolymer mediated dispersion technique, microspheres of PtBA-PCEMA were formed, crosslinked (UV irradiation) and hydrolyzed (trifluoroacetic acid). These dispersible microspheres were approximately 0.5 microns in diameter and contained carboxylic acid-lined channels capable of metal ion uptake [57, 58]. In one case, Cu^{2+} ions were taken up by the nanoporous, water-dispersible microspheres, and evidence for stoichiometric binding by the carboxylic acid residues was presented [57]. Stoichiometric uptake of Pd^{2+} into the nanoporous microspheres was observed in the second example using this approach. Bound Pd^{2+} inside the microspheres was reduced to Pd metal, and the hybrid inorganic/organic nanoporous microspheres were used as dispersible, high surface area catalysts for the hydrogenation of methyl methacrylate [58]. While the authors point out the limitations of their approach relative to more economical routes to porous microspheres, the block copolymer approaches, in general, have the promise of more precise control over pore size, uniformity, and topology. The collective works of Liu and coworkers certainly provide motivation for the technological development of these nanoporous membranes and microspheres.

Using a fascinating self-assembled multi-component system containing a PS-*b*-poly-4-vinylpyridine (PVP) block copolymer and pentadecyl phenol (PDP) [59], Mäki-Ontto et al. reported the formation of nanoporous membranes by selective extraction in 2001 [60]. By mixing PDP and PS-PVP block copolymer, a hierarchical structure consisting of a PS matrix containing hexagonally arranged cylinders composed of PDP complexed to the nitrogen donor on PVP was obtained. In fact, a ("comb-like") layered structure was formed inside the cylinders with the long alkyl chains from the PDP layered with the PVP chains as substantiated by X-ray scattering measurements. Through shearing of these nanocomposites the cylindrical structure was oriented in the direction of flow. By simply subjecting these structures to methanol (a nonsolvent for PS), a significant fraction of the PDP could be removed by dissolution. Using X-ray scattering, the authors observed that the mesostructure was present but evidence for the layered supramolecular assembly within the cylinders was absent. Importantly, macroscopic (mm^3) samples were used in the washing process rather than the more typical thin films (ca. 10–500 nm) used in the nanolithographic applications. IR spectroscopic analysis of the putative mesoporous structure was consistent with the removal of "almost all" of the PDP. The SAXS scattering intensity post washing was also consistent with the formation of pores in the material based on electron density contrast arguments. Unfortunately, no real space images of the samples were given.

In a related publication in 2003, Valkma et al. reported the formation of porous materials in a supramolecular complex containing PS-PVP and an alkylsulfonate Zn(II) complex zinc dodecylbenzenesulfonate [$Zn(DBS)_2$] [61]. A hierarchical structure similar to the PS-PVP:PDP was produced, but in this case a so-called "lamellar-*within*-lamellar" structure was demonstrated. TEM, SAXS and wide-angle X-ray scattering (WAXS) analysis confirmed this structure. Like the structures reported by Lee et al. in 1988 [13], the layered structures in this publication were quite defected. Thus, the authors propose that removal of the $Zn(DBS)_2$ by extraction would not lead to collapse of the resultant porous structure. In fact, subjecting these samples to methanol extraction gave materials with well-defined SAXS reflections consistent with a lamellar morphology. By IR spectroscopy, a "substantial amount" of the $Zn(DBS)_2$ was removed. The TEM images of the extracted materials without stain clearly showed the lamellar morphology and excellent contrast was demonstrated (consistent with pore formation and no pore collapse). By staining with I_2 the remaining PVP became dark in the TEM image, and the contrast inversion was given as evidence for $Zn(DBS)_2$ removal and nanoporosity (Fig. 15). The authors speculate on the exact nature of the pore walls, but all the evidence points to channels that are lined with "uncomplexed" PVP and the possibility of higher functional membranes for biological applications was proposed.

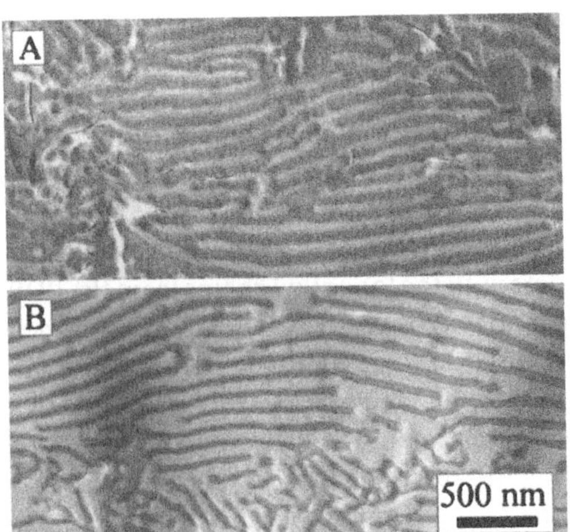

Fig. 15 TEM images of methanol extracted samples of PS-PVP: $Zn(DBS)_2$. **A** unstained and **B** stained with I_2. Reproduced from [61]

Building on the self-assembled structures generated from the PS-PVP:PDP materials described above, Sidorenko et al. developed a related system for the preparation of porous materials [62]. A mixture of a PS-PVP block copolymer and 2-(4′-hydroxybenzene-azo)benzoic acid (HABA) containing a one to one mole ratio of acid to pyridine unit was spin cast as a thin film onto a silicon wafer (and a gold coated silicon wafer). Depending on the casting solvent, a cylindrical morphology was observed either with the cylinder axis parallel or perpendicular to the substrate. The modes of hydrogen bonding between the pyridine repeating units and HABA were implicated in these orientation differences. Spinning films from 1,4-dioxane followed by rinsing with methanol (to remove the HABA) lead to porous thin films with perpendicular pores. If the films are "solvent annealed" with dioxane before methanol treatment, the degree of hexagonal order increases significantly. The authors make the case that the competition for the hydrogen bond acceptors by the solvent is key to the (rapid) reorientation of the films (i.e., from parallel in toluene to perpendicular in 1,4-dioxane). After rinsing the perpendicularly oriented films, the PVP chains can be used to regulate the membrane permeation characteristics in principle. By coating the film from 1,4-dioxane on a gold substrate and rinsing the films with methanol, the channels in the film can be filled with Ni clusters by electrodeposition. Removal of the film by simple dissolution in tetrahydrofuran (THF) resulted in an array of Ni nanodots. This confirms that the channels make a continuous path through the films. This methodology appears to be particularly straightforward for the preparation of nanoporous membranes (Fig. 16).

Lee et al. reported the preparation of nanoporous crystalline sheets of penta-*p*-phenylene using a (oligomeric) block copolymer with a cleavable juncture approach in 2004 [63]. These so-called rod-coil block copolymers of penta-*p*-phenylene and PPO can self-assemble to give layered phases that contain sheets of perforated crystalline penta-*p*-phenylene in which the perforations are filled with PPO [64]. The PPO segment is covalently bound to the

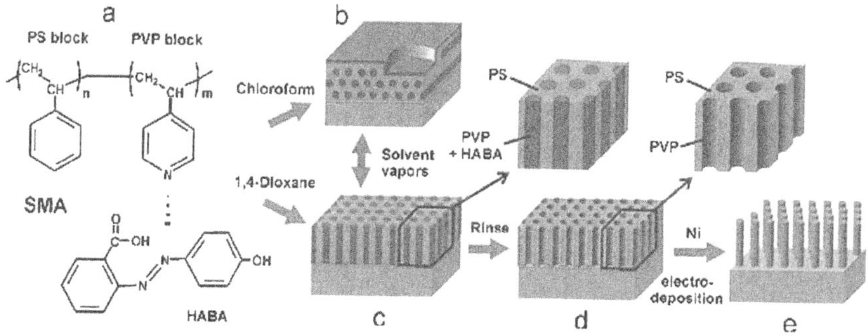

Fig. 16 Preparation of nanoporous membranes from PS-PVP:HABA block copolymer composites. Reproduced from [62]

penta-*p*-phenylene through an ester linkage, and post self-assembly this ester linkage can be chemically cleaved (e.g., by treatment with potassium hydroxide) thus liberating the PPO and leaving behind the crystalline, nanoporous and insoluble penta-*p*-phenylene sheets. Lee et al. showed that this protocol leads to layered and nanoporous "supramolecular crystalline sheets" that contain no PPO (after washing with aqueous HCl and methanol) as confirmed by solid-state ^{13}C NMR spectroscopy. AFM, SAXS, and WAXS analysis corroborated this structural designation and were consistent with 2.1 nm thick crystalline sheets (the length of the penta-*p*-phenylene segment) containing an in-plane nanopore array. The authors of this work went on to show that these nanoporous solids were able to efficiently entrap the hydrophobic dye Nile Red. A hexane solution of Nile Red was mixed with an aqueous dispersion of the nanoporous solid and subjected to ultrasonication. Transfer of the dye from the hexane to the aqueous "subphase" was identified spectroscopically. Favorable interactions between the dye molecules and the hydrophobic interiors of the nonporous penta-*p*-phenylene sheets were implicated.

The ozone etching of a polyisoprene containing triblock copolymer membrane by Hashimoto et al. in 1997 [21] was quoted in a paper on related work by Avgeropolous et al. the following year [65]. In that work the authors prepared silicon-containing triblock copolymers directly related to materials previously prepared (but not structurally characterized) by Hirao et al. [66] and utilized the differential etch rates of the disparate materials to prepare nanoporous membranes from a gyroid forming material. Careful characterization of a poly(pentamethyldisilylstyrene)-*b*-polyisoprene-*b*-poly(pentamethyldisilylstyrene) [P(PMDSS)-PI-P(PMDSS)] triblock by TEM and SAXS showed that the material adopted the gyroid morphology containing two interpenetrating PI networks in a P(PMDSS) matrix. Subjecting that material to ozone resulted in the preferential removal of the PI and a porous structure was generated. Reactive ion etching with O_2 also gave nanoporous materials, but the TEM of those materials was not as clear as in the ozone etching case. By combining ozone etch susceptibility and latent reactivity of the Si-containing monomers (as first reported by Lee et al. in 1988 [13]) the stage was set for the generation of nanoporous materials with both advanced structure and function.

In 1999 Chan et al. reported on the self-assembly and ozone-mediated degradation of two P(PMDSS)-PI-P(PMDSS) triblock copolymers [67]. Both materials adopted the gyroid morphology; one of the triblocks contained PI networks in a P(PMDSS) matrix and the other formed P(PMDSS) networks in a PI matrix. After showing (by TEM) that the gyroid structures were formed in both materials, the authors employed simultaneous ozone treatment and UV irradiation followed by soaking of the thin (700 nm) films in water. This protocol led to the removal of the PI blocks and conversion of the P(PMDSS) to a silicon oxycarbide ceramic. In this way the authors were able to prepare mesoporous ceramics by a relatively mild process. Both nanoporous and so-called "nanorelief" structures were generated depending on which polymer precursor was used. The porous films were characterized by AFM, and the topographical images were compared to computer-simulated images of the putative porous structures. Using homopolymer P(PMDSS) films, the authors also demonstrated that the polymer was converted to an oxycarbide (XPS, ellipsometry, Rutherford backscattering) and that upon thermal treatment post ozone/UV treatment the ceramic films were chemically and thermally stable (400 °C for 1 h). While heat treatment on the porous films was not discussed, the authors suggest that P(PMDSS) struts should be quite strong, and by analogy, thermally resistant. In the concluding paragraph the authors comment that these materials could be used in membrane applications, catalysis, and photonic materials.

Photonic materials are being investigated for their potential optical communication and computation applications, with much focus on the design and preparation of three-dimensional structures [68]. As Urbas et al. pointed out in a 2002 publication [69], "Three-dimensional structures offer the most challenging fabrication problems, and self-assembly can be used to overcome them." In that work, the authors reported on the ozone mediated degradation of a high molecular weight PS-PI block copolymer that forms the bicontinuous gyroid morphology. Calculations on the gyroid structure suggest that it should be a very effective morphology for generating materials with desirable photonic band gaps [70]. By SAXS and TEM the morphology of this 750 kg mol^{-1} material formed after being cast from toluene over two weeks (without further annealing) was confirmed to be gyroid with a cubic ($Ia\bar{3}d$ symmetry) lattice parameter of 258 nm. By exposing a sample of this material to ozone and UV light, the PI was selectively removed leaving a "relief" structure consisting of interpenetrating PS struts. While an image of the material's surface showed a clear porous structure, the middle of the sample was not completely etched (Fig. 17). From reflectivity measurements, both the parent material and the etched material showed photonic behavior. Unfortunately,

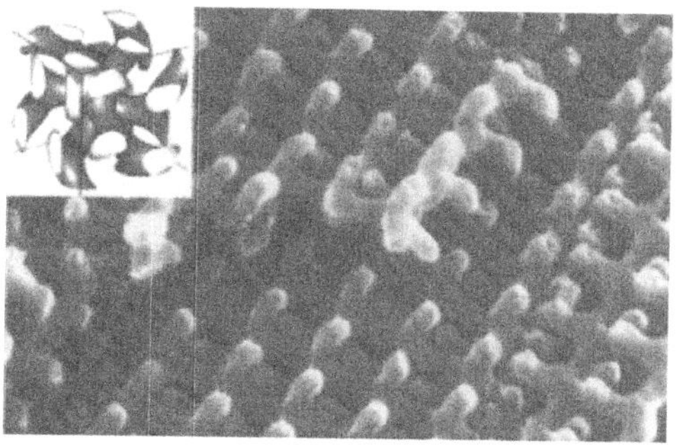

Fig. 17 SEM image of a high molecular weight ozone-etched PS-PI sample that adopted the gyroid morphology. Reproduced from [69]

the dielectric contrast needed for a complete band gap was not achieved in either material. The authors do point to other self-assembled block copolymer structures and filling of the voids with high dielectric constant materials.

5
Monolithic Nanoporous Materials

In most of the nanoporous materials described above, thin films of the block copolymer precursors were employed. This is in large part due to the intended applications of the materials as nanolithographic templates or separation membranes. Furthermore, in many of the examples, the matrix material was crosslinked either during or before the removal of the sacrificial component. In the last example on the generation of photonic materials [69], the ozone mediated degradation of the polyisoprene component was incomplete; although longer etching times were suggested to ameliorate this problem, the ozone, UV, and RIE methodologies may generically suffer from limited degradation in thicker/bulk samples. The development of techniques that allow for the preparation of monolithic nanoporous materials is desirable for many applications. Well-ordered mesoporous silica monoliths for optical and separation applications [71], macroporous polymer monoliths for chromatographic separations [72], and monolithic catalysts for industrial applications are all examples of contemporary uses of monolith materials [73]. Monolithic materials with uniform pore size distributions are also used in bioseparations and as "bioreactors" for biocatalysis applications. Some recent efforts toward monolithic materials with uniform size distributions have been reported by Buchmeiser [74]. In the case of block copolymer template routes, the design and preparation of such nanoporous polymeric monoliths was recently disclosed.

Zalusky et al. published work on the formation of mesoporous polystyrene monoliths from the selective hydrolytic degradation of a PS-PLA diblock copolymer [75]. PS-PLA materials were prepared by a two-step synthesis: anionic polymerization of styrene followed by end capping with ethylene oxide, and polymerization of lactide from the corresponding PS macroinitiator as reported for the preparation of PI-PLA block copolymers in 1999 [76]. The block copolymer contained 40 wt % PLA and formed hexagonally packed cylinders of PLA in a PS matrix. A shear-aligned macroscopic sample of this material was subjected to a water/methanol sodium hydroxide solution above the glass transition temperature (T_g) of PLA but below the T_g of the PS. This resulted in the complete removal of PLA as demonstrated by both NMR spectroscopy and size exclusion chromatography (SEC). The resultant porosity in the material was confirmed by SAXS and SEM. In a simple experiment, the authors demonstrated that the pores in this macroscopic (monolithic) sample were accessible by filling the samples with a colored methanol solution.

This publication established that crosslinking of the polystyrene was not necessary to support the pore structure in monolith nanoporous samples, that mild chemical degradation of an aliphatic polyester is a practical methodology for the generation of bulk porous samples, and that the hydroxyl group derived from the juncture of the PS-PLA material decorated the pore walls of the material.

In a full paper following their 2001 communication, Zalusky et al. described the morphology map for PS-PLA, the tunability of pore sizes, simple methods for cylinder alignment, the type and density of defects in the porous materials, the high temperature stability of the porous polystyrene, and an evaluation of the pore wall functional group accessibility [77]. After establishing the temperature dependence of the Flory–Huggins interaction parameter, χ, for PS-PLA, the morphology map generated revealed no surprising information. Cylinders, lamellae, and the bicontinuous gyroid morphologies were observed as expected for simple coil-coil block copolymers [5]. The cylindrical morphology was observed for many PS-PLA samples that contained between 21 and 43 vol % PLA. The authors reported the use of both reciprocating shear and channel die alignment for the preparation of oriented bulk samples of cylinder-forming PS-PLA materials, and the degree of alignment was characterized by the second order orientation factor. In fact, the channel die processing gave materials with the highest degree of orientation. Good orientation was shown to be important for the generation of bulk porous monoliths that stayed intact during (and after) the basic degradation procedure. The pore fidelity was assessed through evaluation of SEM images acquired along the pore channels (Fig. 18). While defects were observed, the most common defect was two adjacent and parallel cylinders "merging" to form one cylinder (analogous to a "fork in the road"). This type of defect resulted in materials that were able to transport material from one side of the monolith to the other (suggesting there were few "dead ends"). By control-

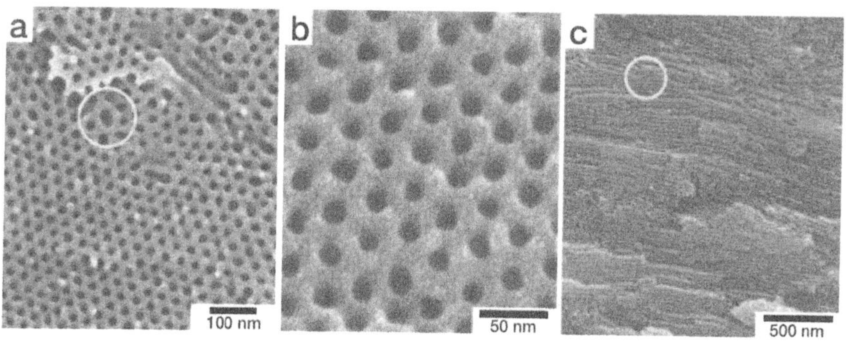

Fig. 18 SEM images of nanoporous PS from monolithic PS-PLA precursors. **a** and **b** perpendicular to the pore axes **c** parallel to the pore axes. Reproduced from [77]

ling the polymer molecular weight, materials with pore diameters from 15 to 45 nm (as determined by SAXS and SEM) were prepared.

In addition to the conventional characterization of the porous materials described above, Zalusky et al. carefully examined their materials by nitrogen adsorption, differential scanning calorimetry and NMR spectroscopy. BET analysis of the N_2 adsorption data was in excellent agreement with the small-angle X-ray scattering and SEM analysis. The specific surface area of the material with 22 nm diameter pores was measured by N_2 adsorption to be 111 $m^2 g^{-1}$. Importantly, the N_2 adsorption data demonstrated that the bulk materials contained an accessible array of pores with narrow pore size distributions, akin to the mesoporous silicas. The pore structure in these mesoporous polystyrene monoliths was stable up to 94 °C, a temperature near the T_g of PS, by small-angle X-ray scattering analysis. Differential scanning calorimetry analysis of these monoliths showed significant exothermic transitions around 95–100 °C. The authors suggested that the collapse of the pore structure at these temperatures leads to "release" of the excess surface energy, and provide a simple calculation consistent with this hypothesis. Finally, the synthesis of PS-PLA and subsequent degradation in the bulk materials should leave a hydroxyl group at the pore surface. The authors calculated an areal density of hydroxyl groups of 1 OH per 4 nm^2 for a sample with 22 nm diameter pores. These hydroxyl groups could be reacted with trifluoroacetic anhydride in the porous monoliths (trifluoro acetic anhydride is a nonsolvent for polystyrene). NMR and IR spectroscopic analysis confirmed that the heterogeneous reaction occurred between the hydroxyl groups and the trifluoroacetic anhydride, thus confirming hydroxyl group accessibility.

In 2004, Ho et al. published a paper on the morphology and degradation of PS-poly-*L*-lactide (PLLA) block copolymers in which the PLLA phase is reported to form "well-organized, hexagonally-packed nanohelices" [78]. The authors argued that the use of a single lactide enantiomer results in the nanohelical structure, and "that the appearance of the nanohelical structure may be attributed to the effect of chirality interacting with the immiscibility of constitute blocks...". While the majority of this work was on the structural characterization of this phase, the authors did manage to generate nanoporous materials by chemical etching (i.e., using an aqueous base as in [77]) as confirmed by SEM.

The nanoporous polystyrene monoliths generated by the degradation of PS-PLA block copolymers have the solubility and thermal characteristics of simple polystyrene homopolymer. The pore structure is lost upon heating above the glass transition temperature or upon exposure to a good solvent (e.g., CH_2Cl_2, THF, or $CHCl_3$). Any potential application of these materials that requires pore structure stability must work within these limitations. The versatility of the polymerization scheme for PS-PLA allows for the generation of other PLA-containing block copolymers and thus other nanoporous matrices. In 2003 Wolf et al. published a paper on the generation of porous polycyclohexylethylene (PCHE) from ordered, cylinder forming PCHE-PLA block copolymers [79]. These authors used a synthetic protocol directly related to that used for the PS-PLA materials, but a hydrogenated version of hydroxyl-terminated PS was employed as the macroinitiator for the lactide polymerization. Several PCHE-PLA samples were prepared and estimates of χ suggested that this pair of polymers was much more incompatible as compared to the PS-PLA couple. Channel die orientation and subsequent hydrolytic degradation of the PLA gave nanoporous monoliths that, unlike their PS cousins, were insoluble in pyridine, ethyl acetate, and acrylic acid. In addition to this improved solvent resistance, the porous structures remained intact after heating up to approximately 135 °C. This superior thermal stability can be directly related to the significantly higher T_g of PCHE as compared to PS (147 vs. 100 °C). This work demonstrated that the methodology for generating mesoporous monoliths as established for PS-PLA was also effective for other matrix materials.

By employing block copolymers with specific functionality at the block juncture, that functionality is by default located at the pore wall interface after degradation of the sacrificial block. In both the PS-PLA and PCHE-PLA materials described by Zalusky et al. [75, 77] and Wolf et al. [79] the hydroxyl group used to form the macroinitiator for the lactide polymerization remains decorating the pore wall. By specific incorporation of other functionality at the block juncture, the chemical groups located on the pore surface can be tuned. Another methodology for controlling the functionality inside the pores using the degradation of PLA approach was demonstrated in 2004 by Mao et al. [80]. In that work the authors blended PS-PLA and PS-PEO block copolymers to form a self-assembled composite consisting of a PS matrix with hexagonally packed cylindrical inclusions consisting of a miscible PLA/PEO "blend". PLA and PEO homopolymers are melt miscible and are both highly incompatible with PS. Thus this AB/AC blending strategy, where B and C are miscible with one another but both immiscible with A, allows for the preparation of porous materials with pore walls "coated" with C if B is degradable and C is not. This is precisely the situation with PLA (B) and PEO (C). Mao et al. showed, for example, that a binary blend of 90% PS-PLA (33 kg mol^{-1}, 30 vol% PLA) and 10% PS-PEO (30 kg mol^{-1}, 45 vol% PEO) forms a hexagonally packed cylindrical microstructure by small-angle X-ray scattering. This blend showed no evidence of PEO crystallinity by differential scanning calorimetry suggesting miscibility with the PLA. In related blends containing even more PEO, no WAXS reflections for crystalline PEO were observed. Furthermore, differential scanning calorimetry analysis for one blend showed two T_gs; one was for the PS phase (100 °C) and a transition at 12 °C was in general agreement with the value predicted using the Fox equation for a miscible blend of PEO and PLA at the appropriate composition. Upon degradation of an oriented (channel die processed) sample of the 90/10 blend with a basic solution, all of the PLA was removed and all the PEO remained in the composite. Both one-dimensional and two-dimensional SAXS data confirmed that both the alignment and microstructural features were preserved. SEM of the degraded sample showed that the monolith generated was indeed porous. Notably, comparison of the "PEO-lined" nanoporous materials and the corresponding PS materials (with the sample pore size) showed very different behavior with respect to water take-up. The parent material without any PEO floated on the surface of water due to its low density. However, a nanoporous piece with as little as 1.2 wt % PEO imbibed water, thus rendering the material denser than water since the density of PS is greater than water (i.e., the nanoporous monolith sank to the bottom of the vial). The PEO coating also rendered these materials much more hydrophilic and unlike the parent PS sample, water was able to wet the pore walls and fill the nanoscopic voids. This new strategy for the preparation of functional nanoporous materials from ordered block copolymers could be quite useful for applications that require the use of aqueous environments (e.g., water filtration, biocatalysis, inorganic templating).

A new method for the bulk chemical etching of ordered block copolymers was published in 2003 by Ndoni et al. [81]. In previous examples of PS-PBD or PS-PI block copolymers that form spherical inclusions of the polydiene, ozone could be used to degrade this polydiene component leaving spherical cavities in the films [22]. While this methodology is effective for thin films, bulk degradation of block copolymers that adopt a spherical morphology had not been demonstrated. That is, even in "accessible" morphologies like the gyroid, ozone-mediated degradations left undegraded material behind [69]. Using a PS-polydimethylsiloxane (PDMS) block copolymer that formed either the bicontinuous gyroid morphology or spherical morphology, Ndoni et al. were able to generate porous materials using an anhydrous HF etch. The byproducts from this etching procedure are volatile and can simply be removed by evacuation. The NMR spectroscopy, SEC, and XPS evidence obtained on the degraded samples were all consistent with near complete removal of the PDMS phase. SAXS measurements showed that (i) the intensity of scattering X-rays increased substantially; and (ii) the ordered state symmetry of the degraded sample was the same as the undegraded materials. Both AFM and SEM images support the formation of nanoporous materials for the spherical samples. While some minor crosslinking of the PS phase was observed, the HF treatment was quite selective for the removal of the PDMS phase. Importantly, by using this procedure, bulk (i.e., not thin film) spherical samples were effectively degraded, thus opening up opportunity for generating interesting bulk closed cell "nanofoams".

In 2004 Yokohama et al. described the preparation of closed nanocellular monoliths using a unique $scCO_2$ processing step applied to a PS-PFMA block copolymer analogous to that used in thin films by the same group (and as described above and in [53]) [82]. Monolithic (100 μm thick films) samples of a PS-PFMA block copolymer were shown by SAXS and surface force microscopy to adopt a spherical domain structure (PFMA spheres in a matrix of PS) with "only short-range order". After exposure of these samples to $scCO_2$ at 60 °C and various pressures (10–30 MPa), the samples were quenched to 0 °C and the carbon dioxide was slowly vented. At high temperature, this process both plasticizes the PS phase and leads to selective segregation of carbon dioxide into the PFMA phase. After the sample is quenched to 0 °C, the continuous PS phase vitrifies, and the PFMA domains remain swollen with CO_2. Careful depressurization leads to closed cell nanofoams as shown in Fig. 19. SEM and complementary SAXS analysis confirmed the structures. The authors reasonably proposed that the PFMA chains are coating the insides of the cells post depressurization. The sizes of the nanocells and therefore cell density can be tuned by the CO_2 pressure utilized in the processing step. "Nanofoams" with cell diameters between 10 and 30 nm, domain spacings between 20 and 40 nm, and cell densities between 1 and 5×10^{16} cm^{-3} were prepared. In one example, an optically transparent foam containing 50% void volume with 30 nm diameter cells and a refractive index of 1.23 was prepared. This is a remarkable achievement, and future applications of these novel materials are anticipated.

In the 2003 publication by Ndoni et al. the authors suggested that using the HF etching procedure on a polydiene-PDMS material would be "highly desirable", and that "Preliminary experiments with polydiene-PDMS block copolymers are showing encouraging results." [81]. About a month after this paper was published two papers were submitted within two weeks of each other (and subsequently published) on the formation of nanoporous materials from crosslinked polydiene materials containing a degradable component. Hansen et al. reported the crosslinking of ordered samples of polyisoprene-*b*-polydimethylsiloxane (PI-PDMS) by dicumyl peroxide [83]. Upon crosslinking the SAXS data suggested that the material was less ordered or even disordered due to the high crosslinking temperature relative to the order-disorder transition and/or because the crosslinking agent was acting as a sol-

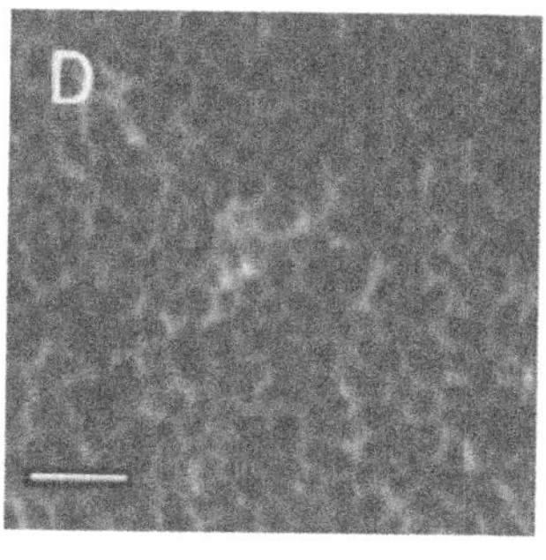

Fig. 19 SEM image of a cryofractured nanocelluar sample of PS-PFMA post CO_2 processing. The scale bar is 100 nm. Reproduced from [82]

vent that reduced the order-disorder transition temperature. After crosslinking, approximately 35% of the PI double bonds remained as determined by solid-state ^{13}C NMR spectroscopy. Crosslinked samples were then subjected to either HF (as in the case with the PS-PDMS samples [81]) or tetrabutyl ammonium fluoride (TBAF). By solid-state ^{13}C NMR spectroscopy, approximately 25% of the PI double bonds remained and complete removal of the PDMS phase was confirmed. In the case of crosslinking at lower temperatures and TBAF degradation, small-angle scattering data consistent with the formation of hexagonally packed cylindrical cavities (weak higher order reflections) were observed. The authors comment that "direct space" imaging will likely corroborate their SAXS scattering data.

Using essentially the same block copolymer, crosslinking agent and degradation media, Cavicchi et al. reported the formation of porous PI monoliths from ordered, aligned, and crosslinked PI-PDMS materials [84]. A blend consisting of 70% PI-PDMS (M_n = 29 kg mol^{-1}, PDI = 1.06) and 30% dicumyl peroxide was aligned at room temperature in a channel die and then heated in an evacuated chamber for 2 h at 150 °C. The crosslinked material was then subjected to a 1.0 M solution of TBAF in THF. Appropriate control samples were also generated. Both gravimetric and infrared spectroscopic analyses were consistent with complete removal of PDMS upon treatment with TBAF. Small-angle X-ray scattering of the crosslinked materials and the degraded materials confirmed that both samples definitely showed oriented, hexagonally packed cylindrical morphologies. SEM images of the materials post TBAF treatment confirmed the porous structures. The authors commented on the large Laplace pressure that exists in nanoporous materials and how that pressure can lead to pore collapse in low modulus materials. By comparing the theoretical limit for pore collapse in crosslinked polyisoprenes, Cavicchi et al. showed that above a critical crosslink density, the nanoporous PI samples were stable. However, at lower crosslink densities, the degraded samples did not show any features in the small-angle X-ray scattering data consistent with pore collapse. Motivated by the work of Cavicchi et al., Muralidharan and Hui developed a model aimed at understanding the origins of and the criteria for stability in nanoporous materials [85]. Briefly, they showed that collapse of an isolated long cylinder depends only on one parameter α, a function of surface free energy, matrix modulus, and pore radius. This work should be considered when designing new nanoporous materials.

Interestingly, the structure of the crosslinked nanoporous PI materials reported by Cavicchi et al. [84] could be reversibly manipulated through solvent swelling. Swelling a porous sample in toluene led to an increase in the domain spacing as compared to the original sample. Removal of the toluene by drying resulted in recovery of the original spacing. An argument based on interfacial tension decreases is given to account for the change. Both of the PI-PDMS examples described above show that with appropriate care, crosslinked elastomeric block copolymers can be degraded to give the corresponding nanoporous rubbers [59]. These have potential as functionalizable porous materials (through the remaining double bonds in the crosslinked PI) [83] and/or as size selective membranes [84].

6 Nanomaterial Applications

Given the variety of methodologies described above for the generation of ordered nanoporous materials from ordered block copolymers, the range of potential applications is quite extensive. By far the most developed application is in nanolithography (Sect. 3), but recently there have been several reports describing interesting and potentially useful application of nanoporous materials prepared by selective removal of the minority component of a block copolymer. For example, in 2003 Misner et al. used a nanoporous thin film generated from the removal of PMMA from an ordered and aligned PS-PMMA block copolymer to sequester CdSe nanoparticles by a straightforward solution deposition process [86]. A film of nanoporous polystyrene with hexagonally packed holes 17 nm in diameter and 29 nm deep was dipped into a heptane solution of tri-*n*-octylphosphine oxide coated CdSe nanoparticles (5 nm in diameter). By capillary action the nanoparticles were swept into the pores (pits) upon removal of the film. The size of the particles, withdrawal rate, and particle concentration all impacted the final composite material. Reasonably, as the concentration of the particles increased so did the average number of particles per pit. At constant particle concentration, the faster the withdrawal, the larger the number of particles found on the surface of the film relative to in the pores. The authors went on to show that using 10 nm diameter particles, most of the pores were filled with only one particle; with smaller particles pores were generally filled with more than one particle (Fig. 20). In all cases the CdSe retained their photoluminescent properties post pore filling. In the concluding paragraph of this publication the authors outlined the salient features and remarkable potential of this simple process for manufacturing complex structures using relatively simple methods. As the authors noted "Directed self-assembly and deposition processes will be necessary for the fabrication of complex devices based on nanoscopic materials".

Kim et al. described an effort toward the fabrication of nanostructured devices utilizing nanoporous materials from ordered block copolymers in 2003 [87]. Using polystyrene films with 20 nm diameter pores (generated from PMMA-PS block copolymers), Kim et al. attempted to replicate the pattern using polydimethylsiloxane as an elastomeric mold [88]. By simply coating a PDMS prepolymer on the nanoporous PS substrate and then crosslinking the PDMS, a negative replica of the nanoporous pattern was generated as confirmed by AFM. The PDMS film had an array of depressions

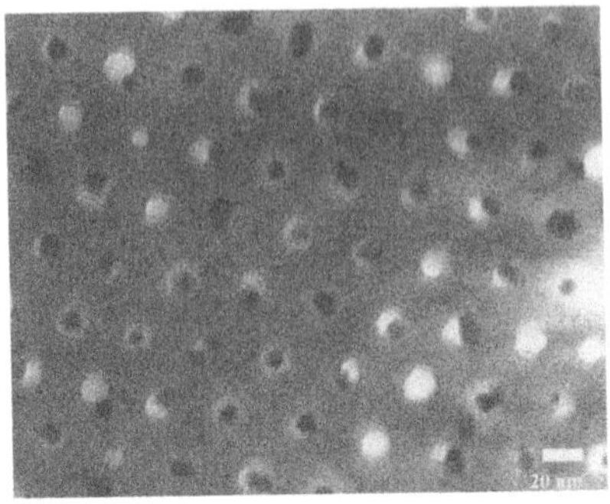

Fig. 20 CdSe nanoparticles residing in the pores of a template prepared from a PS-PMMA block copolymer. Reproduced from [86]

resulting from the inability of the PDMS precursor to wet the polystyrene pores. A simple surface tension argument was used to account for the size and shape of the depressions. Using vacuum crosslinking, a co-solvent, or an electric field, the PDMS prepolymer was effectively drawn into the pores thus producing positive replicas of the nanoporous PS film. However, while the AFM images of the PDMS stamps did show evidence of reproduction, none of these methods resulted in complete filling of the pores or perfect replication. In all cases the heights of the hemispherical caps or discs were only a fraction of the pore depths. Even so, these PDMS replicas may be useful as stamps in so-called "nano-contact" printing.

In 2004, Ha et al. described the use of nanoporous PCHE (generated from PCH-PLA block copolymers [79]) as a nanoconfined crystallization chamber for the selective growth of crystalline polymorphs [89]. Controlling crystal polymorphs is important in, for example, the pharmaceutical industry. By using confined crystallization processes, it may be possible to control polymorphism through simple control of the pore size. A pharmaceutical precursor with six known polymorphs was crystallized from pyridine in the pores of a PCHE monolith containing hexagonally packed 30 nm diameter cylindrical pores. Only one of the crystalline polymorphs was obtained (yellow prisms) along with an amorphous fraction. Heating the monolith to 120 °C afforded only red prisms in the pores of the monoliths. In the absence of the PCHE monoliths, yellow needles are the preferred polymorph. Thus, nanoconfinement in the pores of a nanoporous material generated from an ordered block copolymer appears to control polymorphism. This work was followed by a detailed account of the thermotropic properties of nanocrystals embedded in related nanoporous materials [90]. Simply dissolving the monolith could in principle liberate nanocrystals of an otherwise difficult to obtain polymorph, providing motivation for practical applications.

In another templating methodology, Johnson et al. used both PS and PCHE monoliths as nanotemplates for the generation of nanoscopic CdS

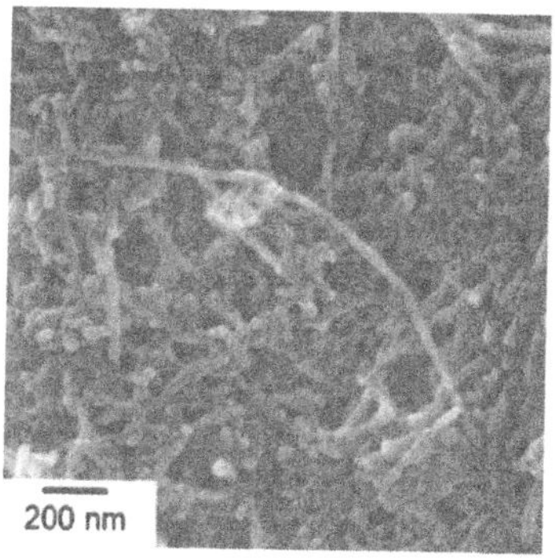

Fig. 21 SEM image of polypyrrole nanowires prepared using a nanoporous PCHE template. Reproduced from [91]

and polypyrrole, respectively [91]. Both nanoporous templates were prepared from the corresponding polylactide-containing block copolymers by established procedures [77, 79]. In the case of the CdS, polystyrene monoliths were first impregnated with cadmium(II) acetate. Upon exposure of the monoliths to a sulfur source, CdS nanoparticles were generated directly inside the pores of the material as evinced by WAXS and TEM. The size of the CdS particles was controlled by the pore size in the monolithic template. The nanoporous structure was not compromised during the synthesis of the CdS particles. Benzene thiol capped CdS could also be prepared inside the porous materials using a related procedure. Nanoporous PCHE monoliths were also used to template the formation of polypyrrole nanowires. In that example, the authors filled the PCHE monoliths with a solution of $FeCl_3$ and after drying exposed the materials to pyrrole vapor. Oxidative polymerization of pyrrole ensued within the nanoporous templates. After removal of the PCHE template, nanowires of polypyrrole were observed by SEM (Fig. 21) and TEM. The diameter of the wires could be controlled by the size of pores in the monolith. These two "proof of principle" nanomaterial template syntheses demonstrate the utility and versatility of nanoporous materials from block copolymer precursors.

7
Summary and Outlook

In this review I have attempted to give a comprehensive overview of the published work on nanoporous materials prepared from ordered block copolymers. Given the numerous successful demonstrations of the block copolymer strategy in the design, synthesis, and applications of nanoporous materials, this methodology holds a great deal of promise for the "bottom-up" engineering of nanostructured materials. While nanolithographic applications predominate the application examples, new developments in chemistry, block copolymer processing, and degradation techniques will lead to a significant increase in the breadth of applications in membrane separations, nanotemplating, and high-surface-area catalysis. It is my opinion that since the work of Nakahama in 1988, progress in the field has certainly been significant, the future will prove equally exciting and new, unforeseen applications of these materials will emerge. While there are many avenues of research in nanoporous materials from block copolymers that remain unexplored, I point out three specific areas that are particularly appealing to me.

The first is in the area of ABC triblock copolymers [92, 93]. These complex materials offer two appealing attributes: morphological complexity and the potential for added functionality. While the four equilibrium morphologies found in diblock copolymers have proven to all be useful precursors to nanoporous material, the morphological zoology of ABC triblocks is far more diverse [94]. This is particularly true with respect to multiply continuous morphologies such as the core-shell-gyroid [95–97], O^{70} [98, 99], and alternating gyroid morphologies [100]. The principles associated with accessing multiply continuous phases in ABC triblock copolymers are now being actively developed, and rules for accessing these phases are emerging [101]. Of course for membrane separation applications, having multiply continuous phases offer the key advantage that orientation of the morphology is not important for the effective preparation of a separation membrane. Furthermore, as pointed out by Urbas et al. [69], there are cubic morphologies adopted by ABC triblock copolymers that would be ideally suited for the preparation of nanoporous photonic materials.

In AB diblock copolymers as nanoporous material precursors, generally one block becomes the matrix and the other is sacrificial. In ABC triblock copolymers the "extra" block can be designed to add functionality. For example, in an ABC core-shell-cylinder morphology with a degradable core (C), the polymer that forms the shell (B) is now coating the pore wall. Depending on the structure of B, specific functionality can be incorporated either in the ABC parent material or be functionalized post degradation of C. My group has made progress in this area using PS-poly(*N*,*N*-dimethylacrylamide)-PLA [102], PS-PI-PLA [103] and PS-PVP-polycaprolactone [104] ABC triblock copolymers. The AB/AC blending strategies described by Mao et al. [80] are related to this approach, with the main idea being to incorporate additional functionality into the nanoporous material.

In the second area, improvements to the thermal and mechanical stability of nanoporous materials from ordered block copolymers should be targeted. To expand the application base for these materials, high temperature stability is a key requirement. For example, in templating applications that require elevated processing temperatures in either thin films or monolithic materials the stability of the pores is of critical importance. In the case of nanoporous PS [77] and PCHE [79] from PS-PLA and PCHE-PLA diblock precursors, the porous structure is not stable once the T_g of the matrix material is approached. Of course crosslinking of the matrix phase will enhance the stability. However, the potential downside of crosslinking strategies is that the matrix material becomes intractable and insoluble. For the template synthesis of nanoscopic materials, simple removal of the template by dissolution facilitates the use of un-crosslinked block copolymers for this application. High T_g matrix materials should be sought after as a remedy for this situation. Also, while no systematic study of the mechanical properties of nanoporous materials from ordered block copolymers has been reported, anecdotally these materials can be quite brittle, especially in the case where relatively unentangled PS is the matrix. In 2004, Lu et al. reported the formation of mechanically robust, porous, and nanostructured polysufone membranes using a PtBA grafting/hydrolysis approach [105]. Later that year, Liu et al. published a paper describing the separation performance of these membranes [106]. Approaches akin to this that are designed to give robust, well-ordered, and uniform nanoporous materials are intriguing and certainly worthy of further exploration. In any event, designing AB diblock copolymers wherein the matrix phase exhibits a higher degree of toughness relative to the typically stiff, but brittle thermoplastics currently employed will be an important undertaking. While many crosslinking chemistries have been developed, the impact on the resultant properties of the nanoporous films has not been thoroughly explored. Interestingly, crosslinking chemistry for the matrix phase based on thermolysis of benzocyclobutane moieties along the backbone of a PS matrix phase with PLA as the degradable component is currently under development for thin film applications [107]. The influence of this crosslinking chemistry on both film and bulk mechanical properties in the resulting nanoporous materials will be revealing.

Finally, while several different degradation techniques have been shown to be effective for preparing nanoporous materials from ordered block copolymer precursors, new degradation techniques should be explored. While in the work of Hedrick et al. [17] thermal degradation of a sacrificial block in microphase separated polyimides has proven to be quite effective for the preparation of low k dielectric materials, the formation of cylinder-forming nanotemplates using a related technique would be difficult given the nature of the step-growth condensation polymerization utilized to prepare these materials. Furthermore, small molecule byproducts generated by thermolysis can plasticize and thus soften the matrix material upon removal, and this can lead to pore collapse [85]. An AB diblock copolymer in which B can be thermally degraded at low temperature and give low-molecular-weight fragments that are both volatile and relatively insoluble in the matrix material is a desirable target. Along these lines, the thermal degradations of PαMS as described by Du et al. [51] and Li et al. [52] are the first published reports of a well-defined self-assembled block copolymer in which the minority component can be cleaved thermally leaving a nanoporous matrix. I believe that this will likely become an important methodology for generating nanoporosity, and extensions of this and related techniques to other systems is anticipated. Other mild degradation techniques based on aqueous etches should also be investigated. By avoiding organic solvents, processing of both thin films and bulk, monolithic materials will become more accessible, practical, and versatile for various applications. In all of the above degradation examples, the methodologies utilized cleave many backbone bonds on the degradable segment. For example, in PLA hydrolysis ester bonds all along the backbone are cleaved, in PMMA degradation by UV irradiation many carbon-carbon backbone bonds are cleaved, and in the ozonolysis of polydienes this is also true. In fact, only the covalent bond that connects the two incompatible blocks needs to be broken. The homopolymer can then be removed by simple dissolution in a good solvent for the cylinder-forming block and a bad solvent for the matrix material. This was nicely demonstrated by Lee et al. in their work with penta-*p*-phenylene as described above [63]. Work in this area has also appeared using a photocleavable anthracene dimer linker between PS and PMMA [108]. Methodologies such as this can facilitate removal of the minority component in nanoporous material precursors [109].

As I hope this review conveys, the design, synthesis, development, and application of nanoporous polymers from ordered block copolymer precursors is an exciting research arena with both fundamental and practical consequences in the fields of block copolymer science and technology.

Acknowledgements First and foremost I thank the students and postdocs in my group that have contributed to our nanoporous material efforts. The enthusiasm, dedication, and productivity of Andrew Zalusky, Roberto-Olayo Valles, Seongho Cho, Johanna Wolf, Huiming Mao, Bret Johnson, Travis Bailey, Shouwu Guo, Pedro Arrechea, and Javid Rzayev are all greatly appreciated. Particular thanks goes to Andrew and Roberto for their important contributions to this review. Volker Abetz, Koji Asakawa, Julie Leiston-Belanger, Guojun Liu, Huiming Mao, Seiichi Nakahama, Christopher Ober, and Thomas Russell are gratefully acknowledged for providing helpful comments and suggestions while this manuscript was in review. The nanoporous materials work in my group has been funded by The University of Minnesota, The National Science Foundation (DMR-0094144), The David and Lucile Packard Foundation, and the Camille and Henry Dreyfus Foundation, and I am grateful for their support.

References

1. Lodge TP (2003) Macromol Chem Phys 204:265
2. Hamley IW (1998) The Physics of Block Copolymers. Oxford University Press, Oxford
3. Hadjichristidis N, Pispas S, Floudas GA (2003) Block Copolymers: Synthetic Strategies, Physical Properties, and Applications. Wiley, New Jersey
4. Hillmyer MA (1999) Curr Opinion Solid State Mater Sci 4:559
5. Bates FS, Fredrickson GH (1990) Ann Rev Phys Chem 41:525

6. Matsen MW, Bates FS (1996) Macromolecules 29:1091
7. Park C, Yoon J, Thomas EL (2003) Polymer 44:6725
8. Hempenius MA, Langeveld-Voss BMW, van Haare JAEH, Janssen RAJ, Sheiko SS, Spatz JP, Möller M, Meijer EW (1998) J Am Chem Soc 120:2798
9. Yamauchi K, Lizotte JR, Hercules DM, Vergne MJ, Long TE (2002) J Am Chem Soc 124:8599
10. Fogg DE, Radzilowski LH, Blanski R, Schrock RR, Thomas EL (1997) Macromolecules 30:417
11. Massey JA, Power KN, Winnik MA, Manners I (1998) Adv Mater 10:1559
12. Beck JS, Vartuli JC, Roth WJ, Leonowicz ME, Kresge CT, Schmitt KD, Chu CT-W, Olson DH, Sheppard EW, McCullen SB, Higgins JB, Schlenker JL (1992) J Am Chem Soc 114:10834
13. Lee J-S, Hirao A, Nakahama S (1988) Macromolecules 21:274
14. Lee J-S, Hirao A, Nakahama S (1989) Macromolecules 22:2602
15. Smith DR, Meier DJ (1992) Polymer 33:3777
16. Hedrick JL, Labadie J, Russell T, Hofer D, Wakharker V (1993) Polymer 34:4717
17. Hedrick JL, Labadie JW, Volksen W, Hilborn JG (1999) Adv Polym Sci 147:61
18. Mansky P, Harrison CK, Chaikin PM, Register RA, Yao N (1996) Appl Phys Lett 68:2586
19. Mansky P, Chaikin P, Thomas EL (1995) J Mater Sci 30:1987
20. Liu G, Ding J, Guo A, Herfort M, Bazzett-Jones D (1997) Macromolecules 30:1851
21. Hashimoto T, Tsutsumi K, Funaki Y (1997) Langmuir 13:6869
22. Park M, Harrison C, Chaikin PM, Register RA, Adamson DH (1997) Science 276:1401
23. Harrison C, Park M, Chaikin PM, Register RA, Adamson DH (1998) J Vac Sci Technol B 16:544
24. Park M, Chaikin PM, Register RA, Adamson DH (2001) Appl Phys Lett 79:257
25. Thurn-Albrecht T, Steiner R, DeRouchey J, Stafford CM, Huang E, Bal M, Tuominen M, Hawker CJ, Russell TP (2000) Adv Mater 12:787
26. Mansky P, Liu Y, Huang E, Russell TP, Hawker C (1997) Science 275:1458
27. Morkved TL, Lu M, Urbas AM, Ehrichs EE, Jaeger HM, Mansky P, Russell TP (1996) Science 273:931
28. Russell TP, Thurn-Albrecht T, Tuominen M, Huang E, Hawker CJ (2000) Macromol Symp 159:77
29. Thurn-Albrecht T, Schotter J, Kästle GA, Emley N, Shibauchi T, Krusin-Elbaum L, Guarini K, Black CT, Tuominen MT, Russell TP (2000) Science 290:2126
30. Shibauchi T, Krusin-Elbaum L, Gignac L, Black CT, Thurn-Albrecht T, Russell TP, Schotter J, Kästle GA, Emley N, Tuominen MT (2001) J Magn Magn Mater 226:1553
31. Kim H-C, Jia X, Stafford CM, Kim DH, McCarty TJ, Tuominen M, Hawker CJ, Russell TP (2001) Adv Mater 13:795
32. Black CT, Guarini KW, Milkove KR, Baker SM, Russell TP, Tuominen MT (2001) Appl Phys Lett 79:409
33. Guarini KW, Black CT, Milkove KR, Sandstrom RL (2001) J Vac Sci Technol B 19:2784
34. Shin K, Leach KA, Goldbach JT, Kim DH, Jho JY, Tuominen M, Hawker CJ, Russell TP (2002) Nano Lett 2:933
35. Guarini KW, Black CT, Zhang Y, Kim H, Sikorski EM, Babich IV (2002) J Vac Sci Technol B 20:2788
36. Jeong E, Galow TH, Schotter J, Bal M, Urache A, Tuominen MT, Stafford CM, Russell TP, Rotello VM (2001) Langmuir 17:6396
37. Bal M, Ursache A, Tuominen MT, Goldbach JT, Russell TP (2002) Appl Phys Lett 81:3479

38. Xu T, Kim H-C, DeRochey J, Seney C, Levesque C, Martin P, Stafford CM, Russell TP (2001) Polymer 42:9091
39. Guarini KW, Black CT, Yeung SHI (2002) Adv Mater 14:1290
40. Jeong U, Kim H-C, Rodriguez RL, Tsai IY, Stafford CM, Kim JK, Hawker CJ, Russell TP (2002) Adv Mater 14:274
41. Jeong U, Ryu DY, Kim JK, Kim DH, Wu X, Russell TP (2003) Macromolecules 36:10126
42. Jeong U, Ryu DY, Kim JK, Kim DH, Russell TP, Hawker CJ (2003) Adv Mater 15:1247
43. Xu T, Stevens J, Villa JA, Goldbach JT, Guarini KW, Black CT, Hawker CJ, Russell TP (2003) Adv Funct Mater 13:698
44. Kim SH, Misner MJ, Xu T, Kimura M, Russell TP (2004) Adv Mater 16:226
45. Kim SH, Misner MJ, Russell TP (2004) Adv Mater 16:2119
46. Asakawa K, Hiraoka T (2002) Jpn J Appl Phys 41:6112
47. Naito K, Hieda H, Sakurai M, Kamata Y, Asakawa K (2002) IEEE Trans Magn 38:1949
48. Asakawa K, Hiraoka T, Hieda H, Sakurai M, Kamata Y, Naito K (2002) J Photopolym Sci Tech 15:465
49. Koji Asakawa, personal communication
50. Olayo-Valles R, Lund MS, Leighton C, Hillmyer MA (2004) J Mater Chem 14:2729
51. Du P, Li M, Douki K, Li X, Garcia CBW, Jain A, Smilgies D-M, Fetters LJ, Gruner SM, Wiesner U, Ober CK (2004) Adv Mater 16:953
52. Li M, Douki K, Goto K, Li Z, Coenjarts C, Smilgies DM, Ober CK (2004) Chem Mater 16:3800
53. Li L, Yokoyama H, Nemoto T, Sugiyama K (2004) Adv Mater 16:1226
54. Liu G, Ding J, Stewart S (1999) Angew Chem Int Ed 38:835
55. Liu G, Ding J, Hashimoto T, Kimishima K, Winnik FM, Nigam S (1999) Chem Mater 11:2233
56. Liu G, Ding J (1998) Adv Mater 10:69
57. Lu Z, Liu G, Liu F (2001) Macromolecules 34:8814
58. Lu Z, Liu G, Philips H, Hill JM, Chang J, Kydd RA (2001) Nano Lett 1:683
59. ten Brinke G, Ikkala O (2004) Chemical Record 4:219
60. Mäki-Ontto R, de Moel K, de Ororico W, Ruokolainen J, Stamm M, ten Brinke G, Ikkala O (2001) Adv Mater 13:117
61. Valkama S, Ruotsalainen T, Kosonen H, Ruokolainen J, Torkkeli M, Serimaa R, ten Brinke G, Ikkala O (2003) Macromolecules 36:3986
62. Sidorenko A, Tokarev I, Minko S, Stamm M (2003) J Am Chem Soc 125:12211
63. Lee M, Park M-H, Oh N-K, Zin W-C, Jung H-T, Yoon D-K (2004) Angew Chem Int Ed 43:6465
64. Lee M, Cho B-K, Zin W-C (2001) Chem Rev 101:3869
65. Avgeropoulos A, Chan VZH, Lee VY, Ngo D, Miller RD, Hadjichristidis N, Thomas EL (1998) Chem Mater 10:2109
66. Hirao A, Nakahama S (1992) Prog Polym Sci 17:283
67. Chan VCH, Hoffman J, Lee VY, Iatrou H, Avgeropoulos A, Hadjichristidis N, Miller RD, Thomas EL (1999) Science 286:1716
68. Norris DG, Vlasov YA (2001) Adv Mater 13:371
69. Urbas AM, Maldovan M, DeRege P, Thomas EL (2002) Adv Mater 14:1850
70. Maldovan M, Urbas AM, Yufa N, Carter WC, Thomas EL (2002) Phys Rev B 65(16):5123
71. Melosh NA, Davidson P, Chmelka BF (2000) J Am Chem Soc 122:823
72. Peters EC, Svec F, Fréchet MJ (1999) Adv Mater 11:1169
73. Boger T, Heibel AK, Sorensen CM (2004) Ind Eng Chem Res 43:4602

74. Buchmeiser MR (2001) Angew Chem Int Ed 40:3795
75. Zalusky AS, Olayo-Valles R, Taylor CJ, Hillmyer MA (2001) J Am Chem Soc 123:1519
76. Schmidt SC, Hillmyer MA (1999) Macromolecules 32:4794
77. Zalusky AS, Olayo-Valles R, Wolf JH, Hillmyer MA (2002) J Am Chem Soc 124:12761
78. Ho R-M, Chiang Y-W, Tsai C-C, Lin C-C, Ko B-T, Hunag B-H (2004) J Am Chem Soc 126:2704
79. Wolf JH, Hillmyer MA (2003) Langmuir 19:6553
80. Mao H, Arrechea PL, Bailey TS, Johnson BJS, Hillmyer MA (2005) Faraday Discuss 128:149
81. Ndoni S, Vigild M, Berg RH (2003) J Am Chem Soc 125:13366
82. Yokohama H, Li L, Nemote T, Sugiyama K (2004) Adv Mater 16:1542
83. Hansen MS, Vigild ME, Berg RH, Ndoni S (2004) Polym Bull 51:403
84. Cavicchi KA, Zalusky AS, Hillmyer MA, Lodge TP (2004) Macromol Rapid Commun 25:704
85. Muralidharan V, Hui C-Y (2004) Macromol Rapid Commun 25:1487
86. Misner MJ, Skaff H, Emrick T, Russell TP (2003) Adv Mater 15:221
87. Kim DH, Lin Z, Kim H-C, Jeong U, Russell TP (2003) Adv Mater 15:811
88. Xia Y, Kim E, Zhao X-M, Rogers JA, Prentiss M, Whitesides GM (1996) Science 273:347
89. Ha J-M, Wolf JH, Hillmyer MA, Ward MD (2004) J Am Chem Soc 126:3382
90. Ha J-M, Hillmyer MA, Ward MD (2005) J Phys Chem B 109:1392
91. Johnson BJS, Wolf JH, Zalusky AS, Hillmyer MA (2004) Chem Mater 16:2909
92. Arai K, Kotaka T, Kitano Y, Yoshimura K (1980) Macromolecules 13:1670
93. Auschra C, Stadler R (1993) Macromolecules 26:2171
94. Bates FS, Frederickson GH (1999) Physics Today 52(2):32
95. Goldacker T, Abetz V (1999) Macromolecules 32:5165
96. Shefelbine TA, Vigild ME, Matsen MW, Hajduk DA, Hillmyer MA, Cussler EL, Bates FS (1999) J Am Chem Soc 121:8457
97. Hückstädt H, Goldacker T, Göpfert A, Abetz V (2000) Macromolecules 33:3757
98. Bailey TS, Hardy CM, Epps TH, Bates FS (2002) Macromolecules 35:7007
99. Epps TH, Cochran EW, Hardy CM, Bailey TS, Waletzko RS, Bates FS (2004) Macromolecules 37:7085
100. Suzuki J, Seki M, Matsushita Y (2000) J Chem Phys 112:4862
101. Epps TH, Cochran EW, Bailey TS, Waletzko RS, Hardy CM, Bates FS (2004) Macromolecules 37:8325
102. Rzayev J, Hillmyer MA (2005) Macromolecules 38:3
103. Cho S-H, Rzayev J, Bailey TS, Hillmyer MA, unpublished results
104. Johnson BJS, Hillmyer MA, unpublished results
105. Lu Z, Liu G, Duncan S (2004) Macromolecules 37:174
106. Liu G, Lu Z, Duncan S (2004) Macromolecules 37:4218
107. Leiston-Belanger JM, Hawker CJ, Russell TP, unpublished results
108. Goldbach JT, Russell TP, Penelle J (2002) Macromolecules 35:4271
109. Goldbach JT, Lavery KA, Penelle J, Russell TP (2004) Macromolecules 37:9639

Adv Polym Sci (2005) 190: 183–226
DOI 10.1007/12_003

Published online: 5 October 2005

Patternable Block Copolymers

Mingqi Li · Christopher A. Coenjarts · Christopher K. Ober (✉)

Department of Materials Science and Engineering, Cornell University, Bard Hall 310, Ithaca, NY 14853, USA
cober@ccmr.cornell.edu

Abstract This article is a review of the chemical and physical nature of patternable block copolymers and their use as templates for functional nanostructures. The patternability of block copolymers, that is, the ability to make complex, arbitrarily shaped submicron structures in block copolymer films, results from both their ability to self-assemble into microdomains, the "bottom-up" approach, and the manipulation of these patterns by a variety of physical and chemical means including "top-down" lithographic techniques. Procedures for achieving long-range control of microdomain pattern orientation as well as the combination of top-down and bottom-up patterning to give multilevel control of block copolymer films are extensively discussed. The level of control over patterning block copolymers that these strategies afford has enabled recent developments in nanofabrication including template nanolithography, template nanoparticle patterning, nanoporous as well as nanoreplicated materials.

Keywords Review · Patternable · Block copolymer · Microdomain · Orientation · Nanofabrication

Abbreviations

1D	One dimensional
2D	Two dimensional
3D	Three dimensional
BA	Benzoic acid
CPS	Close-packed spheres
C	Cylinder
G	Gyroid
L	Lamellae
OTS	Octadecyltrichlorosilane
PαMS	Poly(α-methylstyrene)
P2VP	Poly(2-vinylpyridine)
PAG	Photoacid generator
PBMA	Poly(butylmethacrylate)
PB	Poly(butadiene)
PDMS	Poly(dimethylsiloxane)
PEO	Poly(ethylene oxide)
PETS	Phenylethyltrichlorosilane
PE	Polyethylene
PFS	Poly(ferrocenyl dimethylsilane)
PHS	Poly(4-hydroxystyrene)
PI	Polyisoprene
PMMA	Poly(methylmethacrylate)
PMS	Poly(*para*-methylstyrene)
PS	Polystyrene
SAM	Self-assembled monolayer
SBS	Poly(styrene-*b*-butadiene-*b*-styrene)
SEBS	Poly(styrene-*b*-ethylene butylene-*b*-styrene)
S	Sphere
-*b*-	-Block-
-*r*-	-Random-
f_A	Volume ratio
L_0	Natural period of block copolymers
t	Film thickness
χ	Flory–Huggins parameter
∇T	Temperature gradient

1
Introduction

The trend towards miniaturization of functional devices in modern technology has driven the fabrication and study of nanometer scale periodic structures to such an extent that the new research fields of nanoscience and nanotechnology have emerged [1–3]. Nanoscience and nanotechnology include interdisciplinary research between materials science, engineering, physics, chemistry, and biology. Various techniques have been developed to fabricate functional nanostructures [4], and the use of patternable block copolymers has proved to be one of the most fascinating areas because of its simplicity, elegance, and high productivity [5, 6]. Due to the chemical and physical incompatibility of the different segmental components in block copolymers, these materials tend to self-assemble into a variety of well-ordered nano- and supermolecular structures, with dimensions almost continuously tunable from several to hundreds of nanometers [7–11]. These uniform nanometer-sized patterns are often referred to as "bottom-up" patterns to indicate that they result from the self-assembly of macromolecules.

Alternatively, nanometer-sized patterns can also be fabricated using lithographic methods, which generate features as small as 50 nm. These techniques are often referred to as "top-down" techniques to indicate that the final pattern is formed by the reduction of larger structures. Polymers with block architectures have been investigated for use as microlithographic photoresists since the late 1980s. Most early research focused on polymers having a silicon-containing block [12] because of their improved etch resistance and their superior solubility properties. This early work on these silicon-containing, as well as on fluorine-containing block copolymers, also permitted exploration of unusual developers such as supercritical CO_2. Additionally, block copolymers have been utilized in other top-down processing techniques, such as elastic stamps and "ink" materials for imprint lithography, and imaging resists for scanning probe lithographic processes.

Both top-down and bottom-up patterning approaches have been used in various nanofabrication processes to generate polymers with nanoscopic patterns that can be transferred to other materials. High-resolution nanopatterned substrates, patterned nanoparticles, nanoporous monoliths, as well as nanoreplicated materials have been generated from block copolymer templates. A well-chosen combination of these two patterning approaches provides hierarchical control of block copolymer patterns such that self-assembled nanosized patterns with long-range order and tunable orientation can be generated within lithographically defined submicron patterns.

This review will discuss two types of patterning approaches that can be employed with patternable block copolymers. Due to the fact that most practical applications require block copolymer thin films with large-domain ordered patterns, particular attention is paid to the optimization of bottom-up patterns in block copolymer thin films and the combination of these patterns with top-down approaches to achieve multilevel control of the resulting patterns. Finally, the use of block copolymers as templates to transfer these patterns into other materials for use in nanotechnology is also addressed.

2
Bottom-Up Microdomain Pattern Formation

Block copolymer systems can be categorized in several ways. Depending on their chain configuration, they can be categorized as di-, tri-, star- or multiblock copolymers. By the nature of their self-assembly interactions, they can be additionally categorized as coil–coil, rod–coil, semicrystalline–coil or amphiphilic block copolymers. These various block copolymer systems can self-assemble to form a variety of different microphases and show different phase behaviors in bulk, solution and thin film states. Various aspects of the complex phase behaviors of block copolymers have been presented in previous chapters in this volume by Abetz et al. and Gohy et al., as well as in several other excellent books and reviews [8, 13–15]. As such, only a brief overview of block copolymer phase behavior will be presented in Sect. 2.1 with more emphasis on categorizing different methods to achieve long-range domain alignment in Sect. 2.2.

2.1
Self-Assembled Microdomain Patterns from Block Copolymers

2.1.1
Bulk State Phase Behavior

In the bulk state, the phase behavior of block copolymers composed of flexible coil chains is primarily governed by the mutual repulsion of the chemically incompatible and dissimilar blocks as well as the packing constraints imposed by the connectivity of each block [9, 16]. Coil–coil diblock copolymers are the classic case and extensive theoretical as well as experimental studies have been conducted to predict and exploit their phase behavior. The coil–coil diblock copolymer phase diagram is determined by three independent factors: molecular weight, block composition, and the degree of incompatibility as expressed by the Flory–Huggins parameter, χ. The microdomain size of block copolymers is basically determined by the total molecular weight with resulting morphologies (i.e., microdomain patterns) expected to be uniform-sized spheres, cylinders, and lamellae as well as complex bicontinuous nanostructures (Fig. 1) [16–18]. With the incorporation of an additional building block into the system to form triblock terpolymers, more complex morphologies [19–22] are formed or are predicted, and

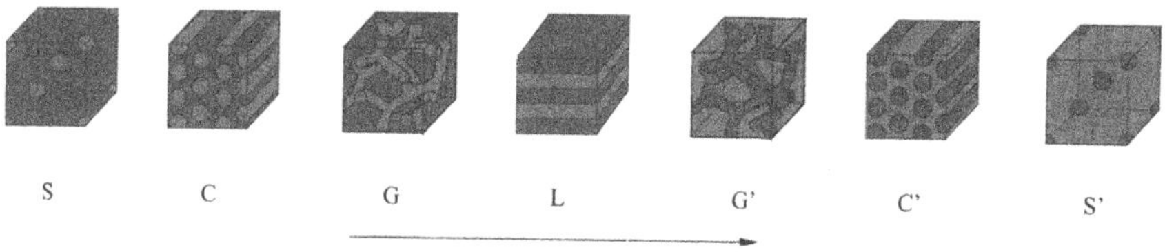

Fig. 1 Mean-field predication of the morphologies for conformationally symmetric diblock melts. Phases are labeled as: S (bcc spheres), C (hexagonal cylinders), G (bicontinuous $Ia\bar{3}d$ cubic), L (lamellar). f_A is the volume fraction

have been best described by Bates, Matsen, and Abetz et al. [17, 23, 24]. The incorporation of further architectural complexity into the system, such as multiblock or star block structures [25, 26], brings more factors into consideration, and thus significantly influences the molecular packing and phase behavior. Further experimental as well as theoretical investigations are still needed to clarify the entire picture of these complex systems.

The bulk phase behavior of block copolymers consisting of flexible polymer coils is remarkably rich. However, by partially or entirely switching a flexible system to one with a rigid rod [10, 11, 27, 28] or with a semicrystalline [29, 30] nature, noncovalent interactions such as hydrophobic and hydrophilic effects, hydrogen bonding, π-stacking, crystallization, electrostatic forces, as well as stiffness asymmetry effects [31] come into play and greatly complicate the details of the molecular packing and the determination of the thermodynamically stable nanostructures. By balancing all of the organizing forces, highly ordered patterns have been extended to a much larger length scale. Unconventional morphologies, such as honeycomb [32–36], mushroom [37], nanoribbon [38], arrowhead, and "wave lamellar" [28] have greatly enriched the block copolymer structure library. Furthermore, the stiff rod segments of these block copolymer systems also endow, or can be coupled afterwards with, electrochemical [33, 36], photophysical [39] or biomimetic [40] functionalities for various nanotechnology applications.

2.1.2
Thin Film State Phase Behavior

In addition to the previously mentioned driving forces that determine the bulk state phase behavior of block copolymers, two additional factors play a role in block copolymer thin films: the surface/interface energies as well as the interplay between the film thickness t and the natural period, L_0, of the bulk microphase-separated structures [14, 41, 42]. Due to these two additional factors, a very sophisticated picture has emerged from the various theoretical and experimental efforts that have been made in order to describe the phase behavior of block copolymer thin films and have been best summarized by Fasolka et al. [14] The picture is further convoluted by the fact that block copolymer nanofabrication processes have usually been performed on thin films with asymmetric compositions (such as 2D cylindrical [43] or 3D spherical [44] microstructures). The extra degrees of freedom inherent to off-symmetric motifs generate very complex systems for both theoretical and experimental scientists. For example, the thin film microdomain motif was found to shift from surface-parallel cylinders to volume-asymmetric lamella [45, 46], due to the interplay between interfacial energies and film thickness. To fully understand and utilize these off-symmetric block copolymer thin films, continued experiments and systematic theoretical calculations are required.

2.1.3
Solution State Phase Behavior

One of the characteristic properties of amphiphilic block copolymers in the solution state is micelle formation in solvents, which are selectively good for one block and poor for another [8, 15, 47, 48]. At polymer concentrations above the critical micelle concentration, block copolymers stabilize the solvent-incompatible blocks by forming a core-shell structure, with the solvent-phobic blocks forming the core to be shielded from solvent, and the solvent-philic blocks forming the corona, which solubilize and stabilize the micelles. In semidilute or concentrated solutions, gelation normally occurs and block copolymer micelles organize into a nanostructure-ordered lyotropic liquid crystal phase [8]. A large variety of block copolymers, including nonionic coil–coil [49], rod–coil [32–36] as well as ionic block copolymers [50], self-organize in aqueous or organic solvent solutions. Well-defined morphologies, such as a 1D lamellar phase, a 2D hexagonal phase of rod-like micelles, a 3D cubic phase of spherical micelles, as well as a 3D bicontinuous cubic phase [47, 48], have been reported and widely used as templates in nanofabrication processes.

2.2
Alignment to Control Long-Range Orientation of Microdomain Patterns

The self-assembly of block polymers, in the bulk, thin film and solution states, produces uniformly sized nanostructured patterns that are very useful for nanofabrication. Optimal utilization of these nanoscopic patterns requires complete spatial and orientational control of the microdomains. However, the microdomains in the bulk state normally have grain sizes in the submicron range and have random orientations. In block copolymer thin films, the natural domain orientations are generally not desirable for nanofabrication. In particular, for composition-asymmetric cylindrical thin films, experimental and theoretical results [46, 51, 52] indicate that microdomains oriented parallel to the substrates are thermodynamically favored when the film surfaces are preferentially wet by either block, while the microdomains oriented perpendicular to the substrates are more desirable as templates for the fabrication of high aspect-ratio nanostructures. Therefore, various strategies have been used for inducing large area orientation of block copolymer microdomain patterns.

In the bulk state, block copolymer long-range orientation is generally achieved by mechanical shear, compression, and elongational fields on the polymer melts or solutions. Steady shear [53–55], capillary extrusion [56], and extensional flow [57, 58] cause certain domain orientations to be favored so that nearly single-crystal-like textures have been produced, as exemplified in the roll-casting process [59]. Besides these mechanical orientation methods, Hashimoto and coworkers [60] have shown that a single, macroscopic, lamellar grain of poly(styrene-*b*-isoprene) (PS-*b*-PI) block copolymer can be achieved using an applied temperature gradient. Due to the sharp temperature gradient ∇T and its slow-moving speed (2 mm/day), the system underwent a "zone-refinement"-like process and a single-crystal-like lamellar grain was grown, with the lamellae normals parallel to the ∇T direction. Amundson et al. [61–63], Böker et al. [64], and Chao et al. [65] have demonstrated that an electric field is another effective means of orienting block copolymer bulk microdomain structures from polymer melts or solutions.

In block copolymer thin films, the perpendicular orientation of microdomains relative to the substrate cannot be achieved by the shear methods developed in the bulk case. Based on the additional variables (film thickness and surface/interface interactions) in block copolymer thin films, as described in Sect. 2.1.2, three different strategies are generally applied for orienting block copolymer thin films:

1. Controlling the substrate physical constraints (topography) and thus the film thickness
2. Chemical modification of the surface to change the substrate–polymer interactions
3. Applying external fields (electric, thermal, eutectic solidification, crystallization, and solvent evaporation etc.) to induce long-range ordering in microdomains with desirable orientation.

Processes which employ combinations of these strategies have proved to be much more effective at yielding uniform long-range ordered patterns than single strategies. Table 1 describes the methods used to control the microdomain orientation of a variety of thin film block copolymers [41, 42, 66–108].

Table 1 Summary of block copolymers studied in order to control microdomain orientations in thin film states

Block copolymer	Structure	Morphology	Orientation method	Ref.
Poly(styrene-*b*-*n*-alkylmethacrylate)	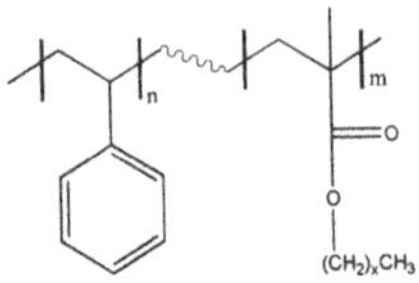	Multiple	Film thickness	[41]
Poly(styrene-*b*-*n* butylmethacrylate)	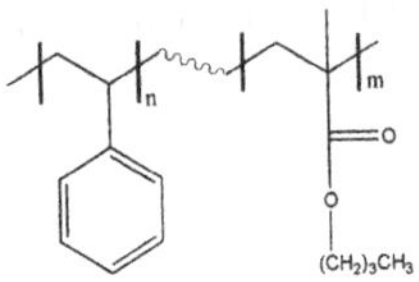	Lamella	Substrate topography	[66]
Poly(styrene-*b*-methylmethacrylate)	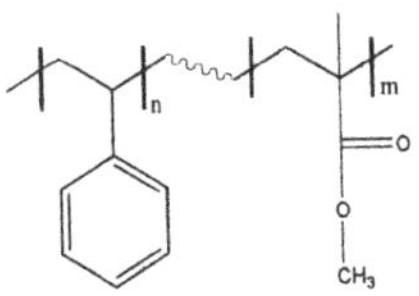	Lamella, cylinder	Neutral surface	[42], [67–71]
Poly(styrene-*b*-methylmethacrylate)	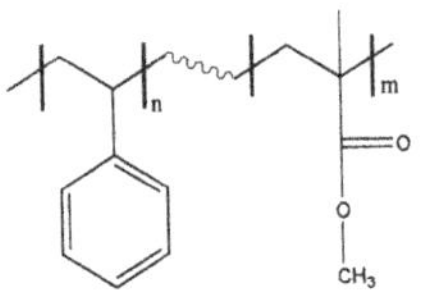	Lamella	Control interfacial interaction	[72]
Poly(styrene-*b*-methylmethacrylate)	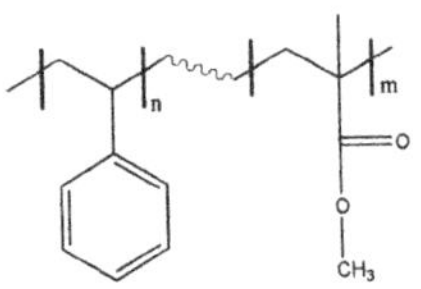	Lamella	Chemical patterned substrate	[73–76]
Poly(styrene-*b*-methylmethacrylate)	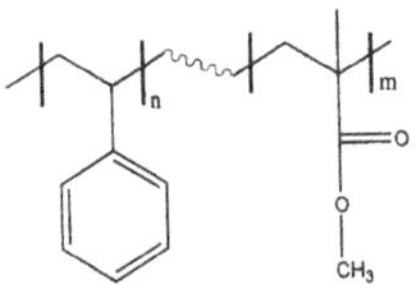	Cylinder	Electric field	[77–81]
Poly(styrene-*b*-methylmethacrylate)	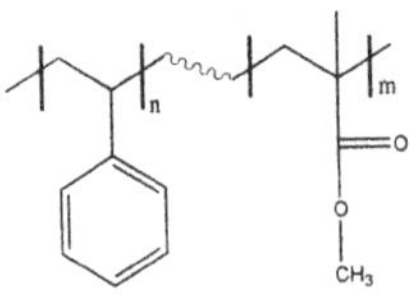	Lamella	Orthogonal electric field	[82]
Poly(styrene-*b*-methylmethacrylate)	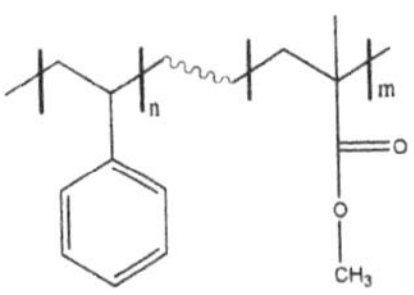	Sphere	Graphoepitaxy	[83]

Table 1 (continued)

Block copolymer	Structure	Morphology	Orientation method	Ref.
Poly(styrene-*b*-methylmethacrylate)	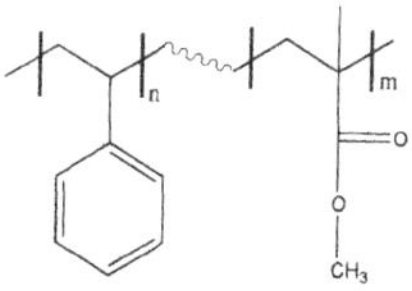	Lamella	Directional crystallization	[84]
Poly(styrene-*b*-2-vinylpyridine)	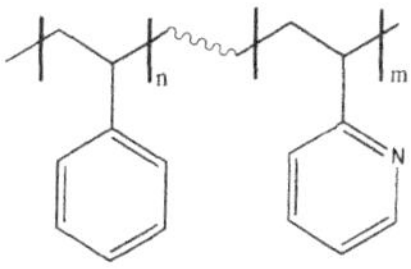	Sphere	Graphoepitaxy	[85–87]
Poly(styrene-*b*-2-vinyl pyridine)	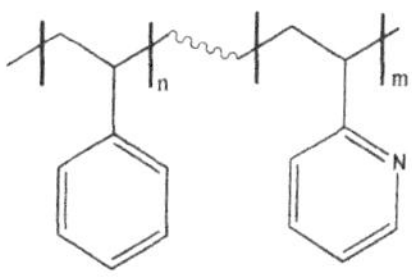	Lamella	Chemical patterned substrate	[88, 89]
Poly(styrene-*b*-ferrocenyldimethyl-silane)	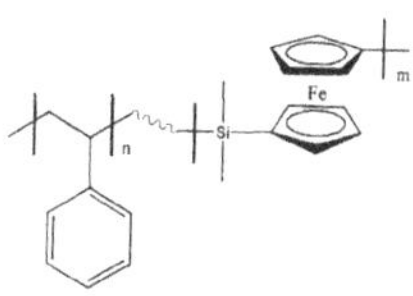	Sphere	Graphoepitaxy	[90, 91]
Poly(ferrocenyldime-thylsilane-*b*-dimethylsiloxane)	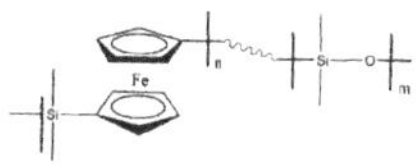	Cylinder	Graphoepitaxy	[92]
Poly(styrene-*b*-ferrocenylethyl-methylsilane)	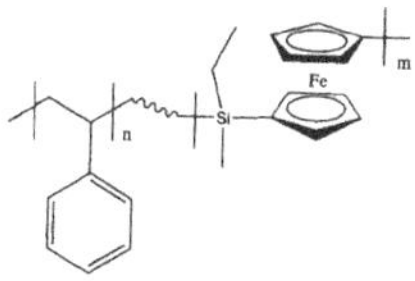	Cylinder	Fast solvent evaporation	[93]
Poly(styrene-*b*-isoprene-b-styrene)	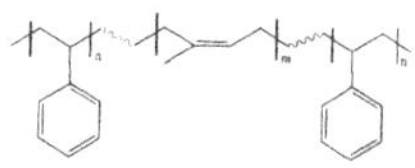	Cylinder	Electric field	[94]
Poly(styrene-*b*-isoprene)	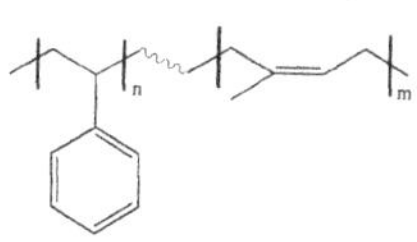	Cylinder	Directional crystallization	[84]
Poly(styrene-*b*-ethylene)	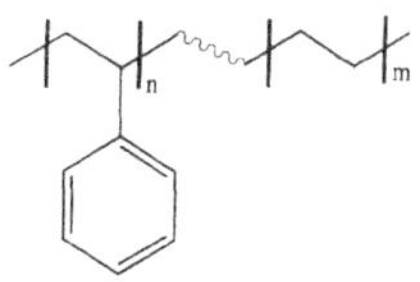	Cylinder	Directional crystallization	[95]

Table 1 (continued)

Block copolymer	Structure	Morphology	Orientation method	Ref.
Poly(butadiene-*b*-ethyleneoxide)		Lamella	Crystallization	[96–98]
Poly(styrene-*b*-butadiene)	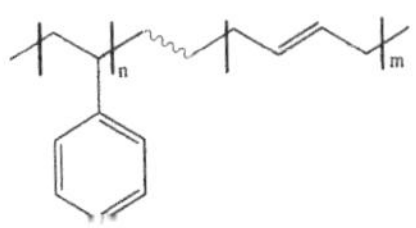	Cylinder	Fast solvent evaporation	[99]
Poly(styrene-*b*-butadiene)	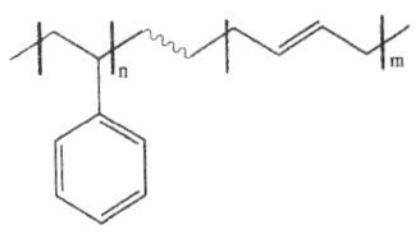	Lamella, cylinder	Fast solvent evaporation	[100]
Poly(styrene-*b*-butadiene)	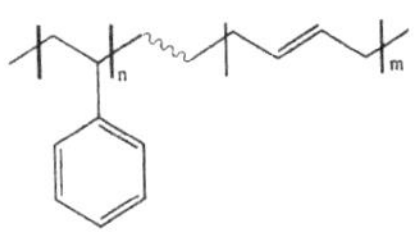	Cylinder	Orthogonal flow field	[101]
Poly(styrene-*b*-isoprene)	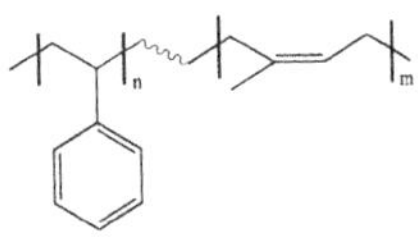	Cylinder	Directional crystallization with Graphoepitaxy	[102]
Poly(styrene-*b*-ethyleneoxide)	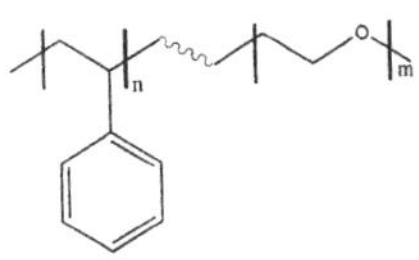	Cylinder	Fast solvent evaporation	[103]
Poly(styrene-*b*-2-vinylpyridine-*b*-*tert*-butylmethacrylate)	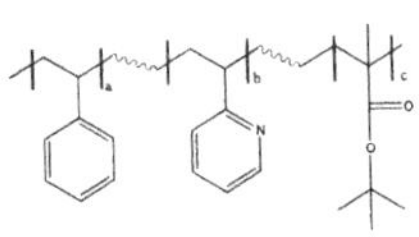	Lamella	Fast solvent evaporation	[104]
Poly(styrene-*b*-butadiene-b-styrene)	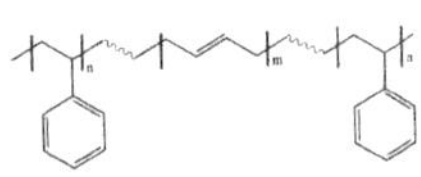	Cylinder	Fast solvent evaporation	[105]
Poly(α-methylstyrene-*b*-4-hydroxystyrene)	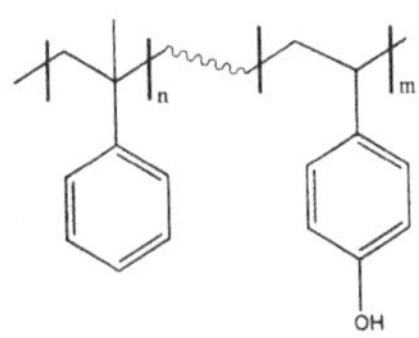	Cylinder	Fast solvent evaporation	[106–108]

2.2.1 Control of Film Thickness

As described in Sect. 2.1.2, block copolymer thin film phase behavior can be controlled by the surface/interface energies as well as the interplay between the film thickness t and polymer natural period L_0. Depending on the nature of the surface interactions, block copolymer thin films can be separated into two general categories [14, 41]:

1. Symmetric boundary conditions [109–111] in which the energies imposed by both film surface/interfaces are identical, such as free-standing films [45] and thin films confined between two identical substrates [67, 111]
2. Asymmetric boundary conditions [41, 109, 110, 113] in which the two surface/interfaces of the thin films have different energies, such as substrate-supported films [66, 68, 114].

Unlike the bulk morphology, block copolymer thin films are often characterized by thickness-dependent highly oriented domains, as a result of surface and interfacial energy minimization [115, 116]. For example, in the simplest composition-symmetric (1D lamellae) coil–coil thin films, the overall trend when $t > L_0$ is for the lamellae to be oriented parallel to the plane of the film [115]. Under symmetric boundary conditions, frustration cannot be avoided if t is not commensurate with L_0 in a confined film and the lamellar period deviates from the bulk value by compressing the chain conformation [117]. Under asymmetric boundary conditions, an incomplete top layer composed of "islands" and "holes" of height L_0 forms as in the incommensurate case [118]. However, it has also been observed that microdomains can reorient such that they are perpendicular to the surface [119], or they can take mixed orientations to relieve the constraint [66].

In the case of $t < L_0$, it has been suggested that perpendicular lamellae are favored in the boundary-symmetric confined film because they avoid the entropic penalty associated with the compressed chain conformations in parallel-oriented microdomains [109]. In boundary-asymmetric substrate-supported films, various kinds of morphologies, including hybrid morphologies that combine surface-parallel and surface-perpendicular components, are predicted, as well as observed, depending on the film thickness difference and surface/interface energies [14, 41, 120].

The use of top-down lithographic techniques to topographically pattern substrates and thereby control the film thickness has been used to create submicron patterns that contain oriented microdomains. This approach is generally described as the graphoepitaxy method and will be discussed in further detail in Sect. 4.1, with other methods which use top-down approaches to control the bottom-up block copolymer patterns.

2.2.2
Control of Substrate–Polymer Interactions

For diblock copolymer films composed of cylindrical or lamellar microdomains, the interfacial interactions dictate the wetting layers at both the substrate and surface interfaces and, consequently, the orientation of the microdomains in the film [41, 45, 67, 109–113, 115, 116]. Therefore, various strategies have been utilized to control the interfacial interactions to achieve large-area microdomains with desirable orientations.

One way of achieving a perpendicular orientation of microdomains is to remove or balance all interfacial interactions, i.e., use nonpreferential or neutral surfaces (Fig. 2). Under these conditions, a perpendicular orientation is favored over a parallel arrangement because of an entropic effect [69, 121]. Genzer et al. [122] and Peters et al. [72] demonstrated control of interfacial interactions by the modification of self-assembling monolayer (SAM) substrates. Symmetric, neutral, and asymmetric wetting of the block copolymer thin films on the modified SAM substrates can be created to direct the block copolymer thin film orientations. Kellogg et al. [67] achieved the same objective by placing random copolymers, consisting of the same monomeric units as the diblock copolymer, at the confining surfaces. In order to avoid potential diffusion of the random copolymers into the confined diblock layer, Mansky et al. [42, 68–71] anchored brushes of random copolymers of PS-*r*-PMMA to silicon substrates, and capped surface-philic perfluorodecanoyl-terminated PS-*r*-PMMA random copolymers at the PS-*b*-PMMA block copolymer/air interface. The compositions of the random copolymers were tailored so that neutral surfaces could be achieved at both substrate and air interfaces. Symmetrical lamellae as well as asymmetrical cylindrical PS-*b*-PMMA block copolymer microdomains were found to orient perpendicular to the interfaces throughout the entire film thickness.

In addition to the aforementioned methods for controlling substrate–polymer interactions uniformly across the entire surface, the use of top-down lithographic techniques to chemically pattern substrates provides spatial control over these substrate–polymer interactions and therefore provides even

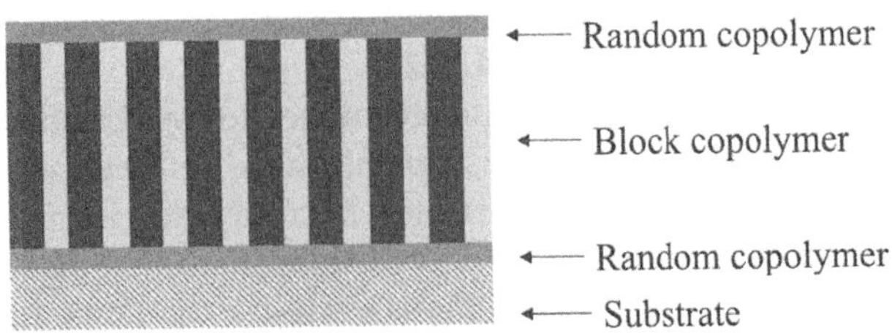

Fig. 2 Schematic representation of control of block copolymer thin film orientations by adjusting polymer–substrate interactions: neutral interfaces

greater control over the microdomain orientations. This approach will be discussed in further detail in Sect. 4.1, with other methods that use top-down approaches to control the bottom-up block copolymer patterns.

2.2.3
Application of an External Field

External fields have been widely used to align block copolymer microdomains. This approach relies on the ability to couple an external applied bias field to some molecular or supermolecular feature, and thus achieve directional control over the microdomains. External fields, such as an electric field [61–63], mechanical flow field [53–55], and temperature gradient [60], have been utilized to control the long-range orientation of block copolymers in the bulk state.

Morkved et al. [77] examined the effect of an electric field on PS-*b*-PMMA block copolymer thin film morphology applied across the film through microfabricated electrodes. In-plane cylinders were found to be aligned parallel to the electric field lines after long annealing times. By adjusting the polymer–electrode interfacial effects, Elhadj et al. [94] also described an approach for producing a metastable state with polystyrene cylinders of poly(styrene-*b*-isoprene-*b*-styrene) (PS-*b*-PI-*b*-PS) triblock copolymers to orient perpendicular to the electric field lines. Thurn-Albrecht et al. [78–81] studied this system in further detail and demonstrated that a vertically ordered cylindrical PMMA microstructure was achieved when the electric field was applied perpendicular to the film surface (Fig. 3). Through theoretical analysis [79, 123], these researchers also proposed that the driving force for the alignment is the orientation-dependent polarizability of the original anisotropic microdomains. Recently, a sequential, orthogonal electrical field approach [82] has been used to gain biaxial control of microdomain pattern alignment, and give long-range, 3D ordered lamellar microdomains in PS-*b*-PMMA thin films.

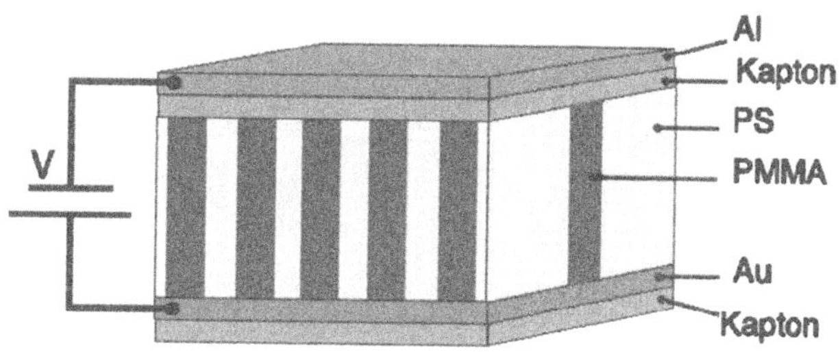

Fig. 3 Schematic representation of control over block copolymer thin film orientation by applying an electric field to orient PS-*b*-PMMA cylinders perpendicular to the substrate (taken from [43])

Application of mechanical stress to a block copolymer thin film can be accomplished by processes associated with crystallization and solvent evaporation. Crystallization is one of the most prominent molecular self-organization processes in nature. Reiter et al. [96] and Hong et al. [97, 98] have explored the possibility of using it to direct microdomain orientation. After spin-casting a symmetrical poly(butadiene-*b*-ethylene oxide) (PB-*b*-PEO) block copolymer onto a substrate, Reiter et al. found that crystallization of the PEO block, combined with a dewetting process, could generate well-aligned vertical lamella on a substrate over macroscopic distances as a consequence of directed growth. De Rosa et al. [84, 95] studied the phenomenon of solvent crystallization-assisted orientation of various block copolymers (Fig. 4). Through eutectic solidification of the block copolymer/solvent mixtures, epitaxial growth of the polymer microdomains was achieved by directional crystallization of a small molecule solvent. Depending on the block copolymer composition, large area surface-perpendicular cylinders or lamella were obtained.

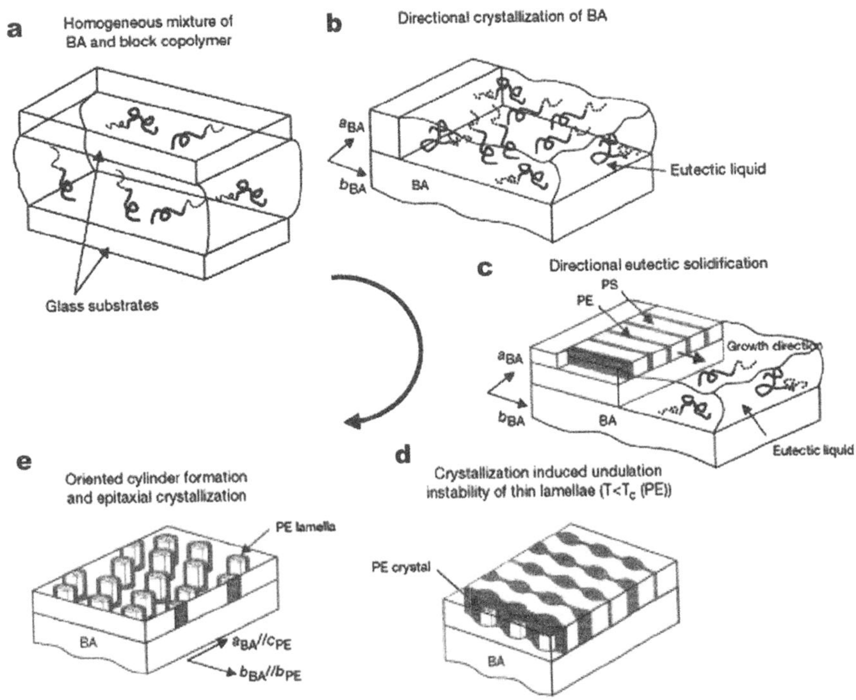

Fig. 4 Schematic representation of control over block copolymer thin film orientation by directional eutectic solidification. The scheme shows control of polystyrene-*b*-polyethylene (PS-*b*-PE) block copolymer microdomain orientation through directional crystallization of benzoic acid (BA). (taken from [95])

Controlling solvent evaporation of cast thin films has proved to be another means of inducing a flow field to align microdomain patterns. The evaporation process is typically highly directional, occurring normal to the surface. Perpendicular orientations of lamella, as well as cylinders, are favored during a fast evaporation process in a variety of block copolymer systems [93, 99, 100, 103–108], possibly due to directional internal strain fields [104] associated with film shrinkage and anisotropic deswelling during the final stages of solvent evaporation. This anisotropic deswelling likely results from the solvent moving through high transport pathways formed by the vertical solvent-philic microdomains to the surface, rather than serial transport through alternating solvent-philic and solvent-phobic layers, which would be present in the parallel-oriented microdomain case [99, 105]. This process generates a zone-refinement-like effect which aligns the microdomains in the vertical direction. Due to the fact that the solvent improves the mobility of the polymer chains and mediates the surface energy difference of the building blocks, which make it feasible to form long range oriented microdomain patterns, nanopatterned surfaces can be rapidly generated by simple spin- or dip-casting processes with the perpendicular orientation of microdomains being possible over a wide range of thicknesses. Recently, by pinning a block copolymer solution droplet on a substrate, Kimura et al. [101] produced two orthogonal flow fields during the evaporation process: a flow field within the droplet directed towards the pinned edge, and the evaporation flow field from the surface to the droplet. Using this approach, ordered arrays of cylindrical microdomains, highly oriented parallel to the surface, were generated over large lateral distances.

3 Top-Down Lithographic Patterns with Block Copolymer Resist Materials

The development of top-down lithographic imaging technologies has been one of the most critical and active areas for the fabrication of new generations of large-area integrated patterns. Polymeric materials have played an important role in lithographic processes by allowing the continued reduction of the minimum feature size through the development of new chemically amplified resist materials that function at shorter exposure wavelengths. Interestingly, the lithographic requirements imposed on polymeric resists, such as sensitivity, adhesion, etch resistance, and solvent developability can rarely be met by simple homopolymers. Copolymers have proved to be useful in this context because they allow the properties of the final material to be tuned by incorporating different monomers together in a variety of ratios.

The most widely used polymeric resist materials are random copolymers, in which monomers with different functionalities are randomly distributed in the polymer matrix so that the final properties are based on the average level of each monomer unit. The improvement of one property often requires sacrificing another property of lithographic merit. The use of block copolymers in which each segment has distinct characteristics and which self-organize into well-defined architectures has provided useful new lithographic properties. It has been shown that the separation of different functional groups of a resist material onto different block segments can offer improvements in dimensional stability, etch behavior, and development characteristics [124].

Besides the isolation of different functionalities, there are additional benefits which accompany the microphase separation of block copolymer resist materials. For example, the efficiency of chemically amplified systems is highly dependant on the ability of photogenerated acid to diffuse within the polymer matrix. Designing block copolymer photoresist materials with polar protecting groups on one block tends to cause the polar photoacid generator (PAG) additives to be distributed selectively in the polar domain. As such, the acid diffusion length required for deprotection is shorter, and the sensitivity of the system is much improved over random copolymers of the same overall composition [125]. Additionally, incorporation of silicon or fluorine-containing units into block copolymer resists causes these low surface energy blocks to migrate and enrich the top surface; thereby possibly acting as a hydrophobic overcoat on the top of the resist and preventing airborne-based contamination. Finally, the possibility of micelle formation with these materials facilitates their development in alternative developers such as supercritical CO_2 which is of interest because of its low environmental impact [126].

In the design of e-beam resists, silicon-containing block copolymers offer many features desirable for high-resolution resist materials, particularly oxygen plasma etch resistance. Hartney et al. [127] first reported the use of poly(*para*-methylstyrene-*b*-dimethylsiloxane) (PMS-*b*-PDMS) as a negative tone e-beam resist. Both DeSimone et al. [128] and Gabor et al. [129] proposed and found that microphase separation and segregation of the silicon-containing block to the air-resist surface not only increases the dimensional stability of the patterns, but also improves the oxygen etch resistance.

In the field of scanning probe lithography, Vasilev et al. [130] have described a process which uses the heat from the probe of an atomic force microscope to provide sufficient thermal energy to drive a highly localized in-situ reorganization of hydrogenated poly(butadiene-*b*-ethylene oxide) (PB-*b*-PEO) mesostructures, giving periodic and nonperiodic hierarchical patterns on the surface. In imprint lithographic processes, thermoplastic SBS and SEBS block copolymer materials have been studied as stamp materials [131], while other block copolymers have been widely used in the solution state as self-assembled "inks" for the prototyping of patterned functional nanostructures. These processes can be thought of as another way of combining top-down and bottom-up patterning approaches, and will be described in further detail in Sect. 4.1.

4 Combination of Top-Down and Bottom-Up Approaches to Give Multilevel Control of Block Copolymer Patterns

Block copolymer nanofabrication can provide large-area periodic functional patterns or objects with feature sizes on the order of tens of nanometers. However, in many practical applications, such as multifunctional on-chip bioseparations, the desired block copolymer microdomains are only required in specific areas on the substrate while the remaining areas on the substrate should be entirely devoid of polymer. In order to satisfy this requirement, a strategy for exerting spatial control over the microdomains is necessary in addition to the usual formation of simple periodic nanostructures. Also, it has been demonstrated that multiaxial guiding forces can generate much improved control of the microstructure order, as compared with a uniaxial aligning force. For these purposes, the convergence of the lithographic top-down approach with block copolymer self-assembly bottom-up nanofabrication within the same system has received attention.

There are generally two strategies for achieving spatial control of nanostructures: first, the substrates are patterned with lithographic techniques to generate top-down patterns, on top of which the block copolymer bottom-up nanopatterns can be subsequently directed; second, the convergence of top-down and bottom- up patterning occurs because the block copolymers contain all of the necessary functionalities to allow both types of patterning. We will describe these two general approaches in the next two subsections and discuss the interesting properties produced by their combination.

4.1 Bottom-Up Block Copolymer Patterns Directed by Top-Down Substrate Patterns

The formation of bottom-up block copolymer patterns within or on top-down substrate patterns is the basis for so-called templated self-assembly processes, in which long-range order and orientation of microdomain patterns can be imposed by a template or "guide". These top-down templates can take a variety of forms including periodic thickness profiles and chemically patterned surfaces.

4.1.1 Topographically Patterned Substrate

A graphoepitaxy method has been developed in which a topographic top-down defined pattern on a substrate is used to direct the epitaxial growth in an overlaying block copolymer bottom-up nanostructure by creating a periodic thickness profile (Fig. 5). Fasolka and coworkers [66] employed a faceted silicon substrate, which has sawtooth-profile corrugations in the nanometer

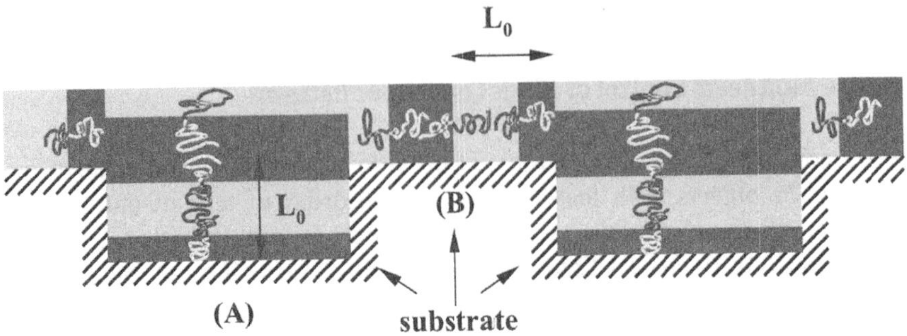

Fig. 5 Schematic cross-sections of thin film morphologies of the topographic pattern grown by a graphoepitaxy method. A micropattern with different lamellar domain orientation is shown. **a** Surface-parallel lamellae, typical of film thickness t greater than the natural equilibrium period L_0. **b** Surface-perpendicular lamellae, typical of film thickness t less than L_0. (adapted from [41])

scale, in the study of symmetric poly(styrene-*b*-butylmethacrylate) (PS-*b*-PBMA) thin films. A flat block copolymer/air surface was observed in the thin film, which implies a periodic thickness profile: thinner above corrugation peaks, thicker above troughs. If the average film thickness is chosen correctly, such that these thickness modulations occur about a critical thickness at which a morphological transition takes place, a lateral microdomain pattern could be developed that mirrors the substrate topology. Through the use of substrate topography, diverse film patterns with a range of length scales have been created, which are consistent with theoretical calculations [41, 132, 133].

Segalman et al. [85] demonstrated the epitaxial growth of a spherical poly(styrene-*b*-2-vinylpyridine) (PS-*b*-P2VP) block copolymer, spin-coated and annealed, on patterned mesa substrates. On a mesa substrate with groove height equal to or greater than the height of one layer of spheres, epitaxial growth of spherical microdomains was nucleated by the edges of the mesas, with the geometry of the edge playing a crucial role in determining the grain orientation and the degree of order in its vicinity [86]. Good long-range order was formed across the entire substrate when the periodic space of the mesa edges was comparable with the grain size. Hahn et al. [87] have also studied the graphoepitaxial deposition of protonated PS-*b*-P2VP micelles on a patterned SiO_2 surface and generated fairly ordered polymer micelle arrays spanning a large area. Cheng et al. [90] applied a similar strategy to the graphoepitaxial growth of poly(styrene-*b*-ferrocenyl dimethylsilane) (PS-*b*-PFS) diblock spheres on patterned silica substrates. With the groove width comparable to the polymer grain size, a nearly perfect alignment of polymer spheres with large area microdomains was formed. Studies of patterned substrate templates with various topographical confinements, in particular with the incommensurability of grooves with an ideal polymer period, have shown that block copolymers behave elastically and can conform to various groove widths, thus leading to arrays with tunable row spacings [91]. In a similar manner, Massey et al. fabricated oriented nanoscopic ceramic lines from cylindrical micelles of a poly(ferrocenyl dimethylsilane-*b*-dimethylsiloxane) (PFS-*b*-PDMS) block copolymer on e-beam defined resist grooves [92], Asakawa et al. utilized 300 nm groove patterns on a polymeric resist film to align the PS-*b*-PMMA microdomains for magnetic data storage applications [83].

In addition to the aforementioned hard substrates, which are normally defined by interference lithography and subsequent chemical etching techniques, Deng et al. applied soft-lithographically defined PDMS molds onto PS-*b*-PFS block copolymer solutions. After the solvent fully evaporated, the PDMS molds were peeled off and submicron patterns containing block copolymer spheres were generated on the substrates [134]. By introducing sol-gel block copolymer solutions into patterned PDMS microchannel molds, through the soft lithographic techniques of micromolding, micromolding in capillaries, and microtransfer molding, Yang et al. [135, 136] have been able to generate patterned mesoscale silica channels that preferred to align along the axes of the PDMS microchannels as directed by the capillary force. These patterned mesoporous silica patterns are potentially useful as waveguide arrays and in integrated optical circuit applications. In a similar manner, Fan et al. [137] demonstrated the utility of sol-gel block copolymers as self-assembled "inks" for three rapid printing procedures: pen lithography, ink-jet printing, and dip-coating of patterned self-assembled monolayers, to form functional, hierarchically organized structures in seconds.

4.1.2 Chemically Patterned Substrate

By modifying the surface/interface energies in a periodic manner, chemically patterned surfaces have also been widely used to control the orientation of nanostructures over large areas (Fig. 6). Theoretical [138–145] and experimental [73–76, 88, 89, 146] results have indicated that with the appropriate surface grating and boundary conditions, lateral control over nanostructures propagates microns away from the surface (deep into the film), thus providing true 3D control of the self-assembly process. Russell and coworkers [73, 74]

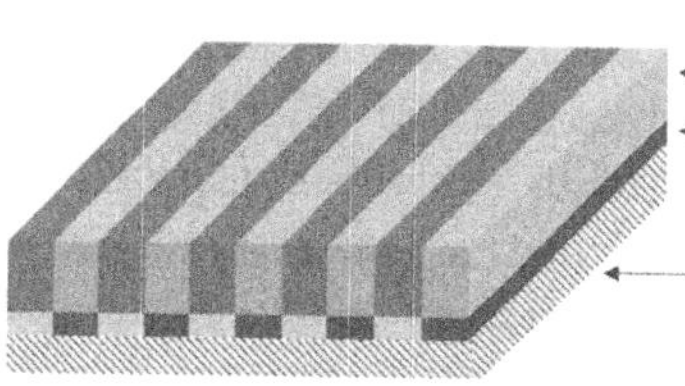

Fig. 6 Schematic representation of controlling the block copolymer thin film orientations by using top-down lithography-defined chemically patterned heterogeneous surface

published the first observation of the orientation of a symmetric PS-*b*-PMMA block copolymer cast on a nanoscopic heterogeneous surface composed of alternating stripes of nonpolar gold and polar silicon oxide. Commensurability of the block copolymer natural period L_0 with the substrate stripe period was found to be essential for production of a surface-directed morphology. When the commensurate condition is fulfilled, the PS and PMMA blocks are attracted to the gold and silicon oxide regions respectively. Lamellar microdomains orient themselves perpendicular to the substrate plane and parallel to the striping. However, a mismatch in length scale of only $\pm 10\%$ is sufficient to cause loss of domain orientation.

Nealey and coworkers [75, 76, 146] took a similar approach and applied lithographically defined self-assembled monolayers as substrates to direct the orientation of block copolymer thin films. After EUV interferometic lithography on octadecyltrichlorosilane (OTS) or phenylethyltrichlorosilane (PETS) monolayers, PS-*b*-PMMA block copolymers were deposited and annealed on the substrates. Due to the selective wetting of PS and PMMA on the unexposed and exposed regions, respectively, they were able to obtain large areas of perpendicular lamella when the commensurate condition was fulfilled.

The use of top-down substrate patterns to direct bottom-up block copolymer patterns has proved to be very efficient in creating oriented hierarchical patterns. However, introduction of yet another structure-directing element to these systems has allowed even more control over the patterning process. Recently, by applying a directional solidification process to a topographically patterned substrate, Thomas and coworkers [102] have been able to apply an additional driving force into the graphoepitaxial growth of the bottom-up block copolymer pattern onto the top-down substrate pattern. By confining a PS-*b*-PI block copolymer thin film between the solvent crystal and the topographic patterned substrate, they have been able to induce two types of PI cylinder domain orientations in the same precise lateral patterns.

4.2
Convergence of Top-Down and Bottom-Up in Block Copolymer Design

The convergence of top-down and bottom-up fabrication in the same block copolymer architecture has been demonstrated using e-beam lithography by Bal et al. [147] and Spatz et al. [148, 149]. Through SEM e-beam exposure of cylindrical PS-*b*-PMMA films, Bal et al. [147] have been able to create cylindrical nanochannels at defined locations from PS-*b*-PMMA on the substrate, and thereby demonstrated its potential to generate integrated magnetoelectronic devices. In this process, the PS matrix is crosslinked in the exposed area, thus making the block copolymer system behave as a negative tone e-beam resist in which the unexposed material is removed in a development step. In a similar manner, by using metal precursor loaded PS-*b*-P2VP block copolymer micellar monolayers as a negative e-beam resist, Spatz et al. created metallic nanodots in microscopically defined locations [148, 149]. In this system, the patternability is proposed to arise from crosslinking of the PS block as well as chemical modification of P2VP/metal salts due to e-beam exposure.

Although e-beam lithography can give excellent spatial control of functional microdomains, this direct-write patterning process is not time-efficient for large-area integration of functional devices. Techniques for rapid patterning of functional nanostructures are thus needed for real-time applications. Ober et al. [106–108] have successfully developed a novel block copolymer

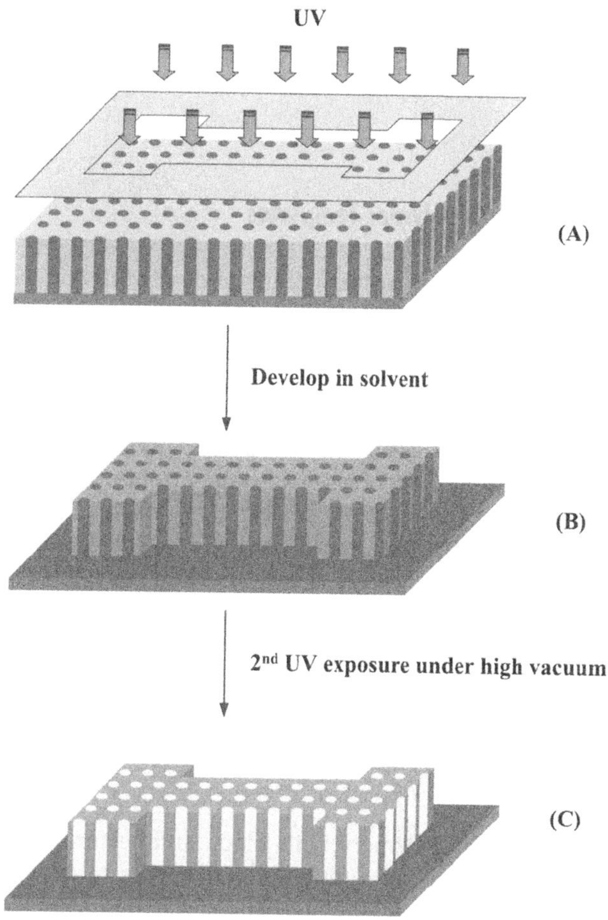

Fig. 7 Novel patternable block copolymers to achieve spatially controlled nanostructures. **a** An asymmetric PαMS-*b*-PHS copolymer/photoacid generator/crosslinker solution was spin-coated on a silicon substrate and formed vertical PαMS cylinders due to rapid solvent evaporation. **b** 248 nm stepper exposure and subsequent development to form micropatterns with features as small as 400 nm. **c** Strong UV irradiation under high vacuum to remove PαMS, thus generating patterned nanochannels

system using poly(α-methylstyrene-*b*-4-hydroxystyrene) (PαMS-*b*-PHS) to achieve spatial control through high-resolution deep UV lithographic processes. Through the incorporation of high-resolution PHS photoresist and PαMS in the block architecture, large-area uniform nanometer-sized cylinders in submicron-sized patterns were generated through simple fabrication processes (Fig. 7). Additionally, this block copolymer was automatically spin-coat aligned with vertical orientations over a wide range of film thicknesses (40 nm – 1 μm), thereby avoiding tedious alignment procedures. Furthermore, this block copolymer system was designed such that the thermodegradable PαMS block could be removed to make a nanoporous material. The transformation of patterns composed of two-block copolymer phases to those of a single block and air, or some other material, are the subject of Sect. 5.3.

5 Patterning with Block Copolymer Templates

All of the material discussed to this point has dealt with patternable block copolymer systems in which a pattern is defined in the polymer, whether by self-assembly of the blocks into microdomain patterns or by lithographically defined patterns. While the process for transferring top-down patterns to the underlying substrates have been well-established, there is also great deal of interest in transferring the self-assembled bottom-up patterns to other materials. Patterning with block copolymers using bottom-up nanostructures as templates in this way opens the door to a variety of nanotechnological applications, which require patterned materials with properties much different to polymeric matrices, such as inorganic networks and metals. High-resolution nanopatterned substrates, patterned nanoparticles, nanoporous monoliths, as well as nanoreplicated materials have been generated from block copolymer templates.

5.1 Templates for Nanolithography

As discussed in Sect. 3, polymers with block architectures have been investigated for use as top-down lithographic materials to exploit a number of their interesting properties. However, as structural dimensions for nanotechnology are pushed to ever-smaller limits, block copolymers have also received attention for use in another new aspect of the lithographic process: as templated resists for the so-called nanolithographic process. Composition-asymmetric diblock copolymer thin films have been shown to be good candidates for use as nanolithographic templates because two-dimensionally large area periodic patterns can be produced on a nanometer-size scale with ei-

Table 2 Summary of block copolymers studied for template nanolithographic applications

Block copolymer	Structure	Morphology	Application	Ref.
Poly(styrene-*b*-butadiene)	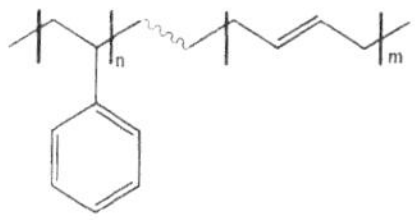	Cylinder	Silicon nitride, germanium nanodots	[44, 152]
Poly(styrene-*b*-isoprene)	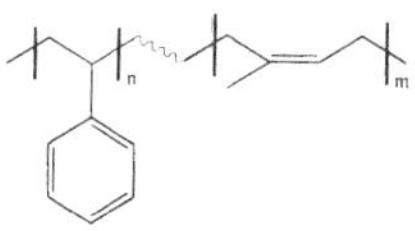	Sphere	Silicon nitride, germanium nanodots	[44, 152]
Poly(styrene-*b*-isoprene)	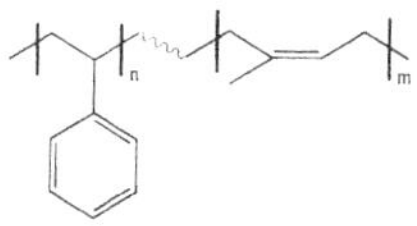	Sphere	Metal nanodots	[154]
Poly(styrene-*b*-isoprene)	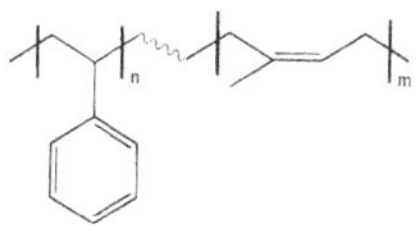	Sphere	GaAs nanodots	[155]
Poly(styrene-*b*-methylmethacrylate)	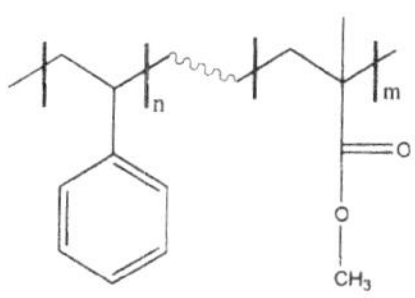	Sphere	$Co_{74}Pt_{26}$ and $Co_{74}Cr_6Pt_{20}$ magnetic media	[83]
Poly(styrene-*b*-methylmethacrylate)	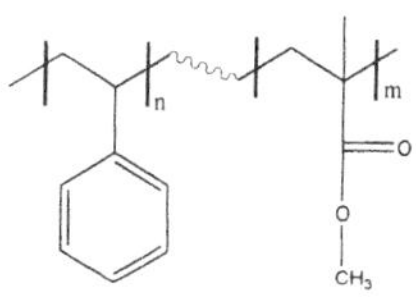	Cylinder	Semiconductor capacitor	[156]
Poly(styrene-*b*-methylmethacrylate)	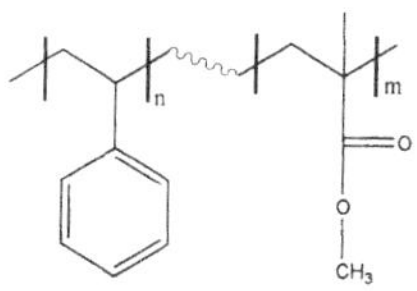	Cylinder	Nanoscopic templates	[157]
Poly(styrene-*b*-2-vinylpyridine)	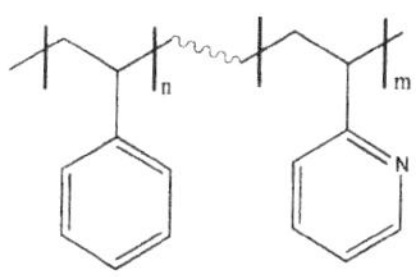	Spherical micelles	GaAs quantum dots	[158]
Poly(styrene-*b*-2-vinylpyridine)	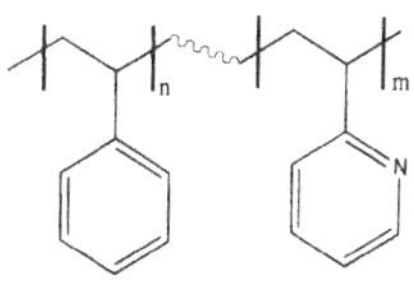	Spherical micelles	Diamond nanocolumns	[158]

Table 2 (continued)

Block copolymer	Structure	Morphology	Application	Ref.
Poly(styrene-*b*-2-vinylpyridine)	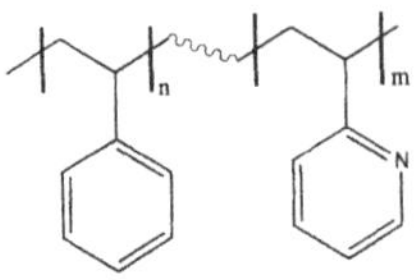	Spherical micelles	Nanoporous gold films	[159]
Poly(styrene-*b*-ferrocenyldimethyl-silane)	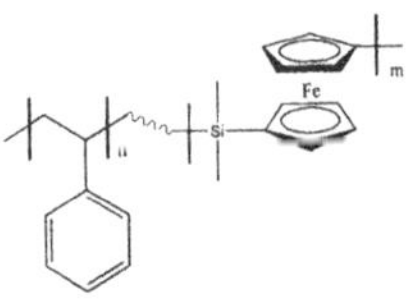	Sphere	Silica nanopillars	[90, 160]
Poly(isoprene-*b*-ferrocenyldimethyl-silane)	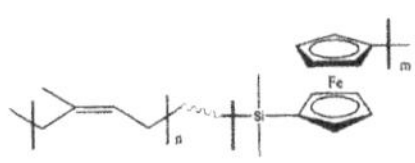	Cylinder	Cobalt magnetic media	[153]

ther a monolayer film of close-packed spheres, or with a thin film containing an array of cylinders oriented perpendicular to the substrate. The periodic nanostructure patterns can be subsequently transferred to various kinds of substrates through a lithographic ion-etching process because of the different etching sensitivities of the two building blocks. The regular nanoscale patterns achieved can be used for a variety of applications, ranging from quantum dot arrays [1, 150], artificial DNA electrophoresis "gels" [44, 151] to high-density magnetic recording media [83, 152, 153]. Table 2 summarizes the various block copolymer systems studied for the template nanolithographic applications [44, 83, 90, 152–160].

Park et al. [44, 102, 152, 155] were the first researchers to demonstrate the template nanolithographic process with well-ordered spherical or cylindrical PS-*b*-PB or PS-*b*-PI diblock copolymers (Fig. 8). Through ozone removal of either the PB or PI minor phase or via selective staining, these block copolymer films were employed as positive and negative resists for a subsequent phase-selective reactive ion etching process. Using this approach, an extremely high-density array ($\sim 10^{11}$ holes/cm^2) of holes and dots were produced on silicon [152], germanium [152], and silicon nitride [44] substrates. By combining block copolymer nanolithography with a multilayer resist technique, they were also able to further demonstrate the generation of high density metal [102] and GaAs [155] nanodots, using the patterned silicon nitride layer as a mask. Using a similar approach, Russell's group [156, 161] and several others [83, 157] have also demonstrated pattern transfer in PS-*b*-PMMA thin films. After UV-degradation of the PMMA minor phase, the nanoporous polymer template structures were successfully transferred to the underlying substrates through a subsequent etching process.

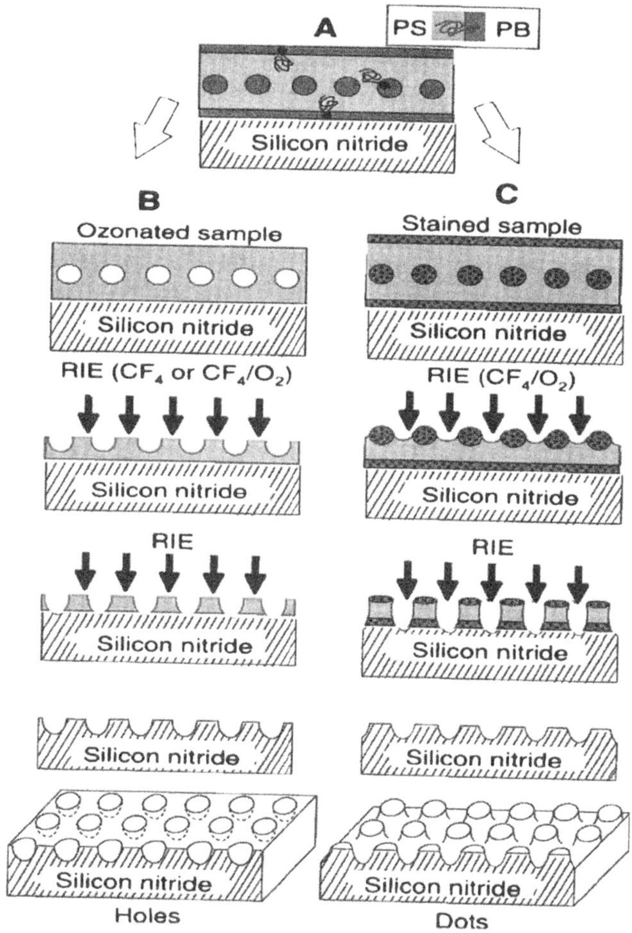

Fig. 8 Schematic representation of block copolymer nanolithography process. **a** Schematic cross-sectional view of a nanolithography template consisting of a uniform monolayer of PB spherical microdomains on silicon nitride. PB wets the air and substrate interfaces. **b** Schematic of the processing flow when an ozonated copolymer film is used as a positive resist, which produces holes in silicon nitride. **c** Schematic of the processing flow when an osmium-stained copolymer film is used as a negative resist, which produces dots in silicon nitride. (taken from [44])

Thomas and coworkers [90, 153, 160, 162, 163] incorporated etch-resistant poly(ferrocenyl dimethylsilane) into block copolymer architectures and used these organic-organometallic block copolymers in a straightforward lithographic process to fabricate functional nanostructures, such as silicon oxide [90] and cobalt magnetic dot arrays [153]. Due to the presence of iron and silicon atoms in the blocks, the nanodomains of the organometallic portion were resistant to removal by reactive ion etching. In this way, high etch selectivity between the different building blocks and thus high aspect-ratio hierarchical patterns were achieved.

Spatz et al. [158, 159, 164, 165] used an approach based on the self-assembly and formation of block copolymer mono-micellar films as masks for template nanolithography. By either loading gold ions on the P2VP core of poly(styrene-*b*-2-vinylpyridine) (PS-*b*-P2VP) micelles [158, 159, 165], or by selective growth of Ti on top of the PS domains [164], highly selective etch contrast was achieved for fabrication of nanostructures on a variety of substrates. By sputtering gold onto the template lithography patterned GaAs substrate and subsequent solvent relief, they have also been able to generate nanoporous gold films with uniform pore sizes [159].

5.2 Patterning Nanoparticles

The well-defined microdomains of block copolymer systems can themselves serve as devices, for example as 1D photonic crystals [166] and stimuli-sensitive materials [167], or they can provide nanocompartments for fabricating spatially defined functional nanoparticles or nanoclusters. In general, depending on the sequence of reactant loading and reactor formation, two approaches have been used in these patterning processes: First, nanoparticles or precursors are phase-selectively aggregated on a microphase-separated block copolymer nanostructure to produce well-defined nanoparticle patterns along the bottom-up block copolymer patterns. Second, functional precursors are incorporated into the polymer by pre-attachment to one of the block copolymer building components, or are homogeneously mixed with the block copolymer. The subsequent self-assembly process, accompanied by the in-situ selective phase separation, produced functional nanoparticles within one of the block copolymer microdomains; thus being patterned after subsequent pyrolysis or sequestration. Table 3 summarizes the various block copolymer systems studied for nanoreactor applications [93, 158, 159, 164, 168–181].

The selective incorporation of functional nanoparticles within well-defined block copolymer bottom-up patterns has been well-studied and numerous examples appear in the literature. Due to the preferential wetting of one of the copolymer blocks by metals, Morkved et al. [168, 169] demonstrated the selective aggregation of a variety of types of metal atoms along highly anisotropic PS-*b*-PMMA and PS-*b*-P2VP block copolymer microdomains. Kane et al. [178], Brown et al. [172] and Sone et al. [182] have independently produced PbS nanoclusters through a procedure involving the selective sequencing of metal ions into the block copolymer microdomains and subsequent reduction with H_2S gas. In a similar approach, several groups have reported the selective sequestration of organometallic complexes into block copolymer microdomains or mi-

Table 3 Summary of block copolymers studied for patterning nanoparticle applications

Block copolymer	Structure	Morphology	Application	Ref.
Poly(styrene-*b*-methylmethacrylate)	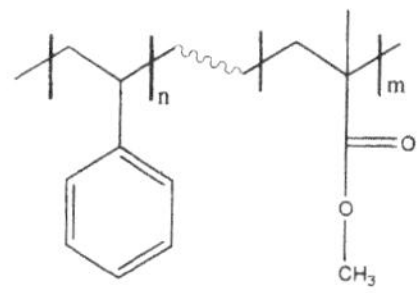	Lamella	Gold islands	[168, 169]
Poly(styrene-*b*-methylmethacrylate)	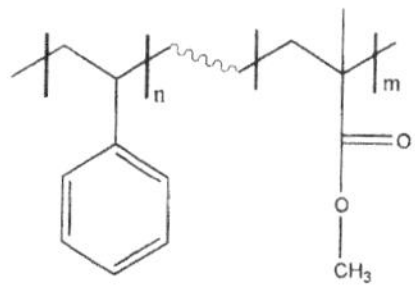	Lamella	Co nanocluster	[170, 171]
Poly(styrene-*b*-*tert*-butylacrylate)	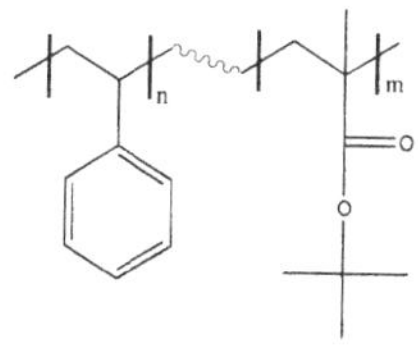	Sphere	Pt, Ag, or PbS nanoclusters	[172]
Poly(styrene-*b*-2-vinylpyridine)	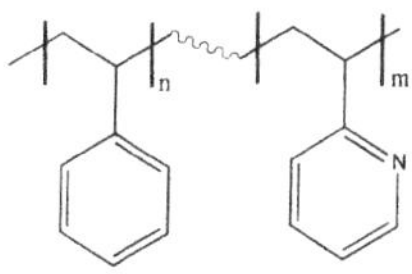	Sphere	Gold islands	[168]
Poly(styrene-*b*-2-vinylpyridine)	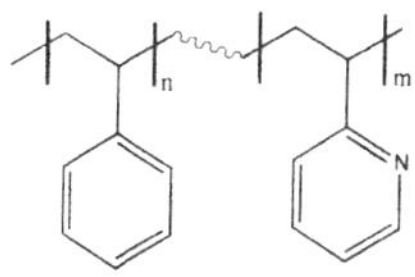	Lamella	Fe, Fe–Co, Co–Ni nanoparticles	[173, 174]
Poly(styrene-*b*-2-vinylpyridine)	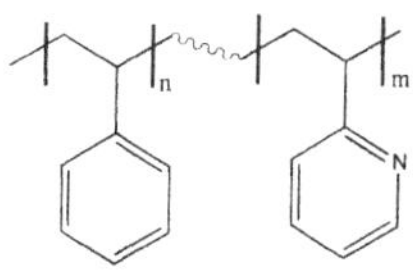	Spherical micelle	Gold nanoparticles	[158, 159], [164], [175, 176]
Poly(styrene-*b*-butadiene-*b*-styrene)	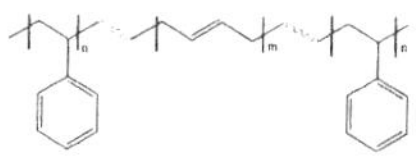	Cylinder	Nanopattering of BaTi	[177]
Poly(methyltetracylododecene-*b*-substituted-2-norbornene)	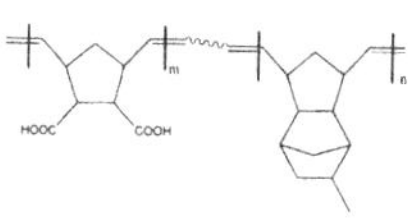	Sphere and wormlike	PbS, CdS nanoclusters	[178, 179]

celles to generated a variety of metal, inorganic, or conducting polymer nanoparticles with hierarchical patterns [158, 159, 164, 165, 175–177, 180].

Table 3 (continued)

Block copolymer	Structure	Morphology	Application	Ref.
Poly(styrene-*b*-ferrocenylethyl-methylsilane)	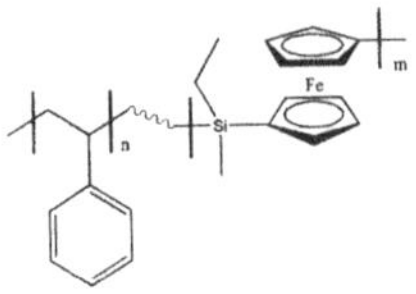	Cylinder	magnetic α–Fe_2O_3 nanoparticles	[93]
Poly(styrene-*b*-isoprene-*b*-styrene)	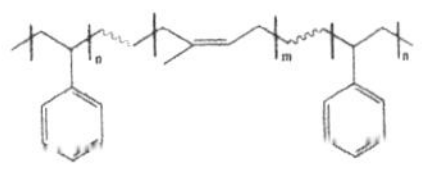	Lamella	Gold nanoparticles	[180]
Lysine-*b*-cysteine diblock copolypeptide	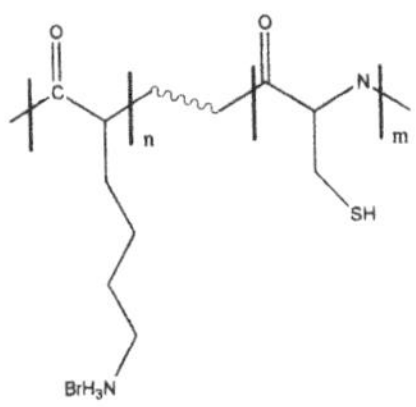	Sphere	Silica, gold nanoparticles	[181]

There have been numerous reports concerning the incorporation of functional precursors into block copolymer structures [93, 153, 160, 162, 163, 179, 183]. For example, poly(ferrocene)s have been successfully introduced into block copolymer architectures by Vancso et al. and Manners et al. [90, 153, 160, 162, 163, 184, 185]. After self-assembly processes and subsequent pyrolysis, ceramic materials that contain magnetic α– Fe_2O_3 nanoparticles have been generated [93]. Iron, iron-cobalt, and cobalt-nickel alloy nanoparticles have been generated within the P2VP domains of PS-*b*-P2VP block copolymers by first making homogenous solutions of organometallic complexes [173, 174]. A variety of types of nanoparticles or nanoclusters have been cooperatively assembled into block copolymer nanoreactive domains using similar approaches [170, 171, 181, 186].

5.3 Nanoporous Pattern Formation

Nanoporous materials have been generated using many different approaches [187–193]. Currently, there is intense interest in the topological control of porosity in inorganic and organic nanoporous materials for applications in catalysis [194, 195], separations [196–198], dielectric [199, 200] and optical communications [136, 201]. Materials with a defined porosity in the nanometer scale are more highly desirable for many high-tech applications [202] than materials with a pore size distribution. Owing to their ability to form well-defined nanostructures, block copolymers have received attention as templates and matrices for fabricating well-defined nanoporous or mesoporous materials. The nanofabrication processes can be roughly categorized into three general approaches (Fig. 9):

1. Organic block copolymer micelles serve as structure-directing reagents in a sol-gel mixture with inorganic precursors to form mesoporous inorganic structures
2. The minor phase of block copolymer microdomains are selectively removed by either chemical or physical degradation, thereby generating a defined porosity in crosslinked polymer bulk matrices or thin films
3. Microporous structures form on cast film from rod–coil block copolymer micelles

Table 4 summarizes the various block copolymer systems studied as nanoporous and mesoporous materials [32, 33, 35, 44, 79, 106–108, 152, 154, 155, 193, 199, 203–219].

There have been numerous studies using surfactant and amphiphilic block copolymer micelles to fabricate mesoporous inorganic materials and several excellent reviews appear in the literature [48, 220–223]. Substantial processing advantages have been demonstrated using amphiphilic block copolymers as structure-directing reagents instead of low molecular weight surfactants, such as continuously tunable properties by adjusting block copolymer composition and molecular weight, as well as larger mesoscopic order. The nanofabrication procedure usually starts with a self-assembled amphiphilic block copolymer solution under acidic conditions. Inorganic reagents are subsequently added into the system and the reaction product forms a solid phase after solvent evaporation, mixed with the self-organized block copolymer template. The removal of template materials is generally achieved by thermal calcinations [203], extraction (by conventional solvent [204], or by supercritical fluid [205]) and chemical decomposition [206]. Mesoporous inorganic materials, with uniform pore size and high degrees of orientational order, are thus generated. Different amphiphilic block copolymers [203, 207–210], in particular poly(alkylene oxide) containing diblock [209, 224] or triblock copolymer systems [136, 199, 200, 203, 207, 208, 225], have been used to fabricate mesoporous silicate [199, 200, 203, 209, 210] or oxides [207, 208] with randomly distributed [199, 200] or oriented nanopores [203, 209, 210]. Recently, in order to improve the structural order by avoiding simultaneous precursor condensation and structural evolution, a process was described to generate highly organized 3D mesoporous silicate [226]. This process resembles biomineralization, in which the poly(alkylene oxide) block copolymer template forms first and is followed by the infusion, selective condensation of the precursor, and removal of template.

Synthetic strategies are also directly applied to block copolymer systems to generate nanopores in polymer matrices or thin films by selective chemical or physical degradation and removal of minority organic polymer do-

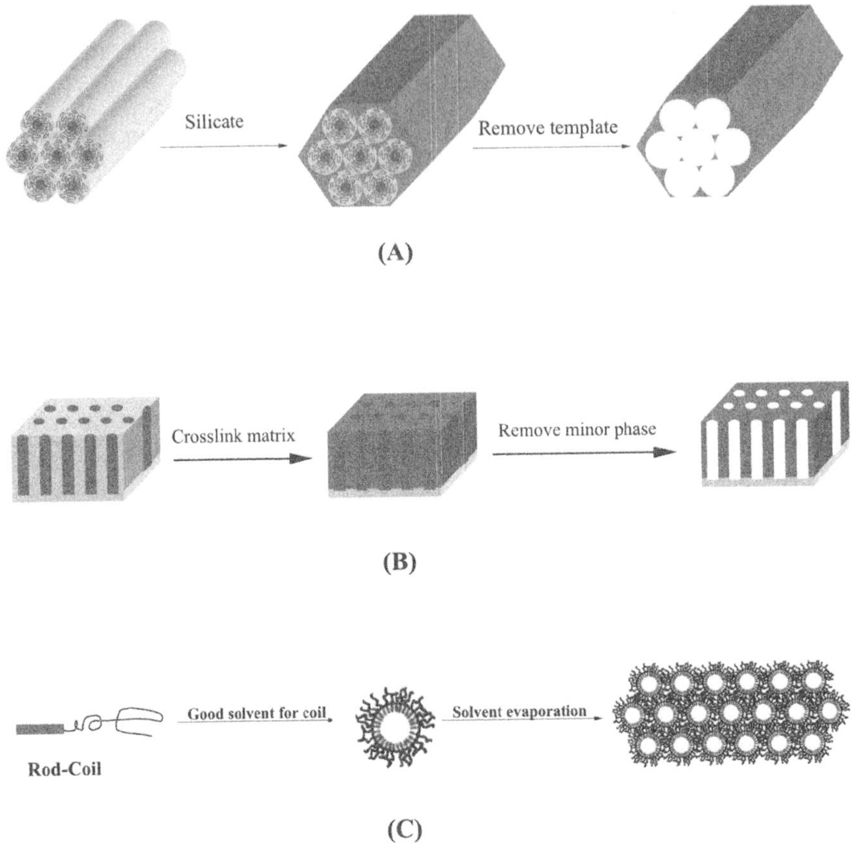

Fig. 9 Schematic representation of three approaches to generate nanoporous and mesoporous materials with block copolymers. **a** Block copolymer micelle templating for mesoporous inorganic materials. Block copolymer micelles form a hexagonal array. Silicate species then occupy the spaces between the cylinders. The final removal of micelle template leaves hollow cylinders. **b** Block copolymer matrix for nanoporous materials. Block copolymers form hexagonal cylinder phase in bulk or thin film state. Subsequent crosslinking fixes the matrix; hollow channels are generated by removing the minor phase. **c** Rod-coil block copolymer for microporous materials. Solution-cast micellar films consisted of multilayers of hexagonally ordered arrays of spherical holes. (Adapted from [33])

mains. The optimum nanofabrication procedure usually starts with the self-organization of a block copolymer in the bulk or thin film state. Shear [216–218], electric fields [78–81, 123] or other means [107] are then applied to align the microstructure, thus producing the desired orientation within long-range ordered microdomains. Subsequent pore generation generally involves crosslinking, thereby fixing the larger polymer block domains, as well as partial [215, 219, 227], or complete degradation [44, 78–81, 107, 123, 152, 154, 155,

Table 4 Summary of block copolymers studied for nanoporous and mesoporous applications

Block copolymer	Structure	Morphology	Application	Ref.
Poly(ethyleneoxide-*b*-propyleneoxide-*b*-ethyleneoxide)	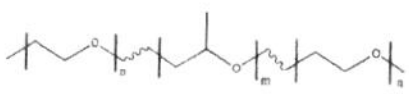	Hexagonal micelles	Mesoporous silica and oxides	[203–207]
Poly(ethyleneoxide-*b*-propyleneoxide-*b*-ethyleneoxide)	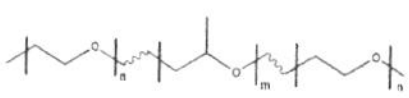	Spherical micelles	Nanopores for low *k* dielectric applications	[199]
Poly(ethyleneoxide-*b*-propyleneoxide-*b*-ethyleneoxide)	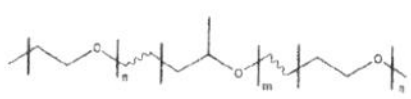	Hexagonal micelles	Mesoporous metal oxides	[208]
Poly(isoprene-*b*-ethyleneoxide)	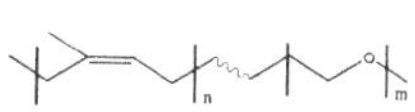	Multiple	Mesoporous aluminosilicate	[209]
Poly(butadiene-*b*-vinylpyridinium)	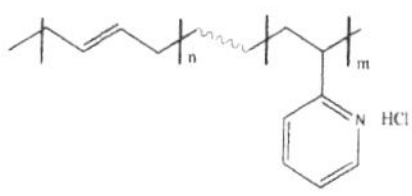	Spherical micelles	Mesoporous silica	[210]
Poly(styrene-*b*-butadiene)	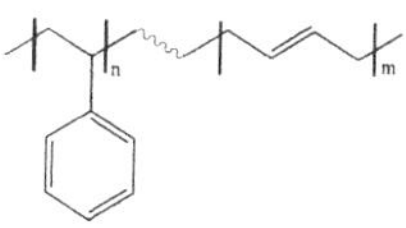	Cylinder	Nanoporous polymer	[44, 152]
Poly(styrene-*b*-isoprene)	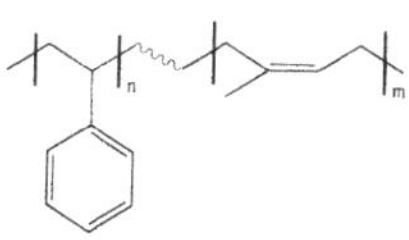	Sphere	Nanoporous polymer	[154, 155]
Poly(4-vinylphenyldimethyl-2-propoxysilane-*b*-isoprene-*b*-4-vinylphenyldimethyl-2-propoxysilane)	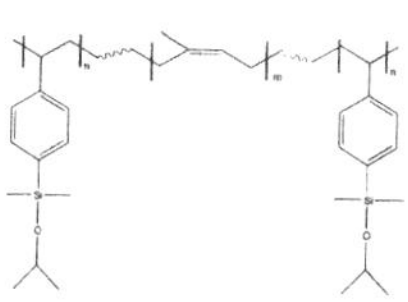	Lamella, sphere/cylinder	Nanoporous polymer	[211]
Poly(styrene-*b*-methylmethacrylate)	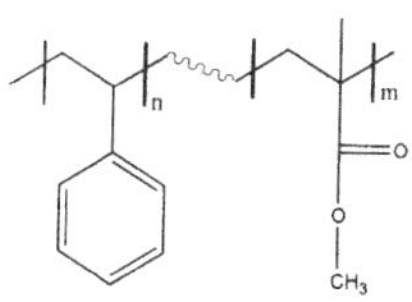	Cylinder	Nanoporous polymer	[79], [212–214]
Poly(*tert*-butylacrylate-*b*-cinnamoylethylmethacrylate)	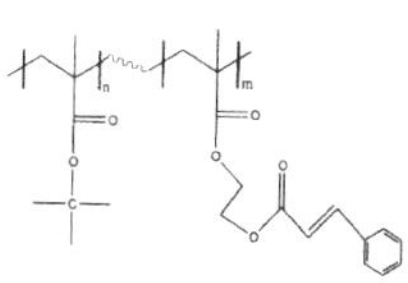	Cylinders	Nanoporous polymer	[215]

Table 4 (continued)

Block copolymer	Structure	Morphology	Application	Ref.
Poly(isoprene-*b*-pentamethyldisilyl-styrene-b-isoprene)	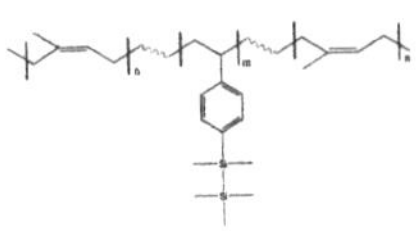	Double gyroid	Nanoporous polymer	[193]
Poly(styrene-*b*-lactide)	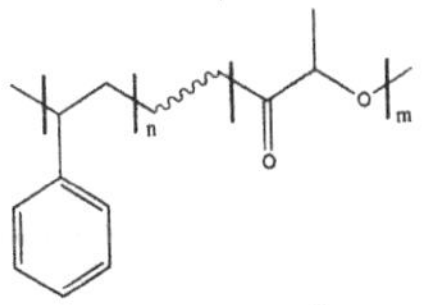	Cylinder	Nanoporous polymer	[216, 217]
Poly(cyclohexyl-ethylene-*b*-lactide)	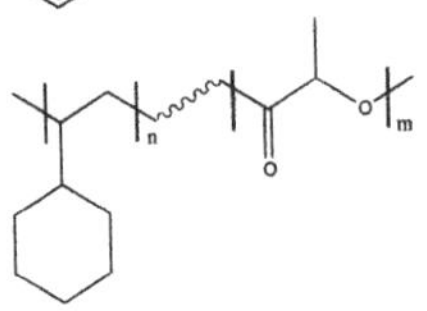	Cylinder	Nanoporous polymer	[218]
Poly(α-methylstyrene-*b*-4-hydroxystyrene)	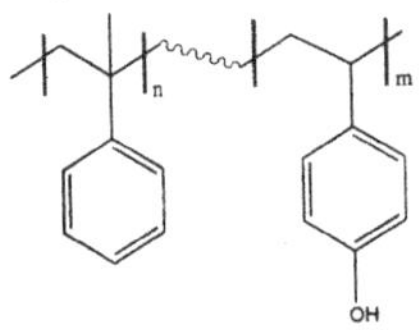	Cylinder	Nanoporous polymer	[106–108]
Pentadecyl phenol modified poly(styrene-*b*-4-vinylpyridine)	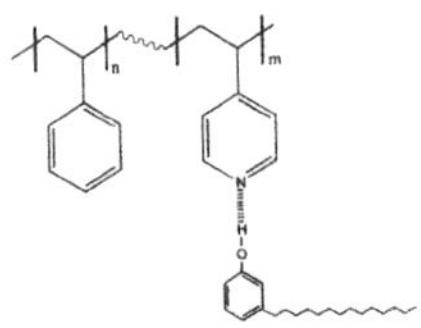	Cylinder	Nanoporous polymer	[219]
Poly(phenylquino-line-*b*-styrene)	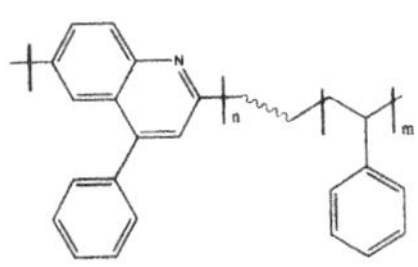	Spherical micelle	Microporous materials	[33]
Poly(2,5-dioctyloxy-*para*-phenylene vinylene-*b*-styrene)	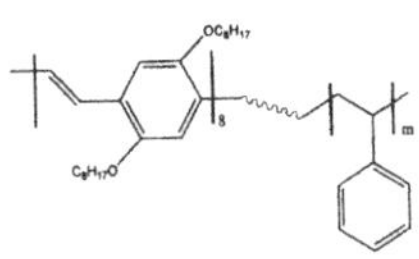	Spherical micelle	Microporous materials	[35]
Poly(styrene-*b*-paraphenylene)	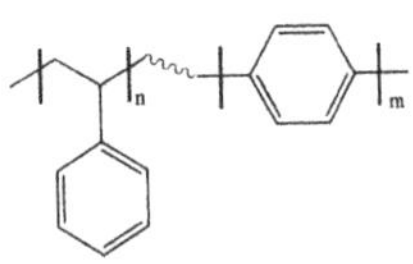	Spherical micelle	Microporous materials	[32]

193, 211, 216–218] of the smaller block domains. Because of the uniform domain size and mild removal conditions, precise control can be achieved to yield well-defined monolithic materials [228].

Numerous examples on the preparation of nanoporous polymers in the aforementioned manner appear in the literature. Nakahama et al. made the first attempt with a siloxane-functionalized PS-*b*-PI system [211]. Upon forming a cylindrical packing array of polyisoprene cylinders, a monolithic polymer thin film was formed by subsequent crosslinking through siloxane moieties and ozonolysis to remove the PI phase. After this groundbreaking work, various block copolymer systems, including poly(styrene-*b*-butadiene) [44, 152], poly(styrene-*b*-isoprene) [154, 155], poly(styrene-*b*-methylmethacrylate) [78–81, 123], poly(isoprene-*b*-pentamethyldisilylstyrene-*b*-isoprene) [193], poly(*tert*-butylacrylate-*b*-2-cinnamoylethylmethacrylate) and related systems [215, 227], poly(styrene-*b*-lactide) and related systems [216–218], pentadecyl phenol modified poly(styrene-*b*-4-vinylpyridine) [219], poly(α-methylstyrene-*b*-isoprene) [107] as well as poly(α-methylstyrene-*b*-4-hydroxystyrene) [106–108], have been employed to create ordered monolithic polymers in bulk and thin film states. Additionally, physical means, such as volume contraction triggered by crosslinking [213], solvent-induced surface reconstruction [212], and extraction of homopolymer in a blend with block copolymer matrix [214, 229], have also been employed to create nanoporous polymer films.

The self-assembly of rod–coil block copolymers cast from a selective solvent causes the formation of microporous films with a honey-comb morphology [32–36]. These films consist of periodic micrometer sized pores in a hexagonal lattice. Recent studies have revealed that the formation of such structures are due to the condensation of small water droplets on the surface of the polymer solution during the rapid evaporation of the organic solvent [11, 230, 231]. The polymers prevent the coalescence of the water droplet and vitrify the pattern created by the condensation of the water droplet during the evaporation of organic solvent. Although the rod–coil type architecture seems to have little effect on the structure formation, such materials still generate great interest for a variety of applications, such as membrane [232], pattern template [35], and photonic bandgap materials [33].

5.4 Nanoreplicated Pattern Formation

One of the most important applications of nanoporous membranes is as "nanoscaffolds" in template synthesis, to replicate the structural features of the nanopores, or patterns, into metals [233], carbons [234], semiconductors [235, 236], conductive polymers [237, 238], and other materials [239]. The important characteristics of template synthesis have been best reviewed by Martin [188, 240]. In short, it is a robust, general method suitable for the replication of nanoscale features into almost all kinds of materials, which can be retrieved afterwards without structural damage. Table 5 summarizes the various block copolymer systems studied for nanoreplication applications [35, 43, 154, 155, 159, 233, 235, 236, 239, 241].

The Russell group has applied the template synthesis approach to nanoporous films generated from UV-treated PS-*b*-PMMA copolymers [43, 147, 233, 235, 241], which were pre-aligned perpendicular to the substrate by an electric field. Through direct current electrodeposition, they fabricated high-density vertical arrays of ferromagnetic cobalt nanowires (Fig. 10a) [43]. Through subsequent work, they also demonstrated the successful replication

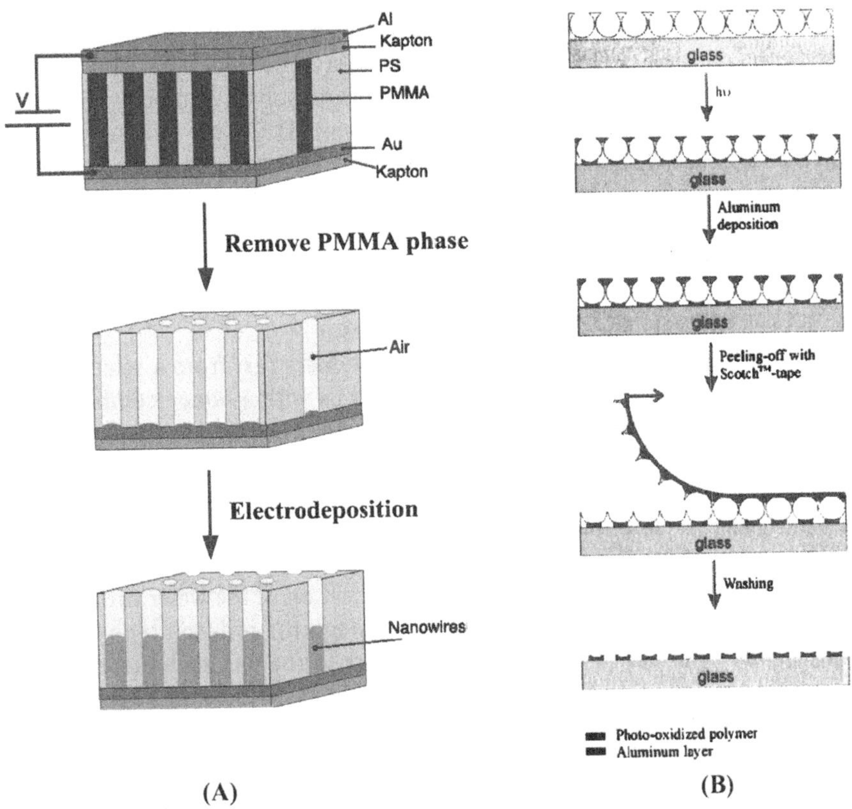

Fig. 10 Schematic representation of the nanoreplication processes from block copolymers. **a** Growth of high-density nanowires from a nanoporous block copolymer thin film. An asymmetric PS-*b*-PMMA diblock copolymer was aligned to form vertical PMMA cylinders under an electric field. After removal of the PMMA minor component, a nanoporous film is formed. By electrodeposition, an array of nanowires can be replicated in the porous template (adapted from [43]). **b** Hexagonally packed array of aluminum caps generated from rod-coil microporous structures. Deposition of aluminum was achieved on the photooxidized area of the rod-coil honeycomb structure (Taken from [35])

Table 5 Summary of block copolymers studied for nanoreplication applications

Block copolymer	Structure	Morphology	Application	Ref.
Poly(styrene-*b*-methylmethacrylate)	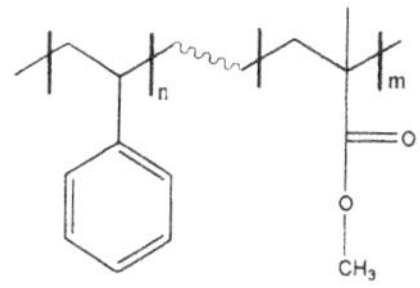	Cylinder	Ferromagnetic nanowires	[43]
Poly(styrene-*b*-methylmethacrylate)	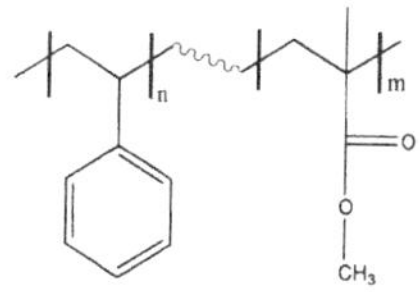	Cylinder	Silicon dioxide nanoposts	[235]
Poly(styrene-*b*-methylmethacrylate)	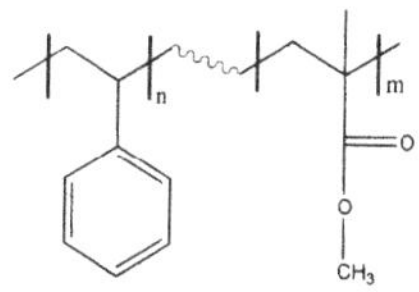	Cylinder	Metal nanodots	[233]
Poly(styrene-*b*-methylmethacrylate)	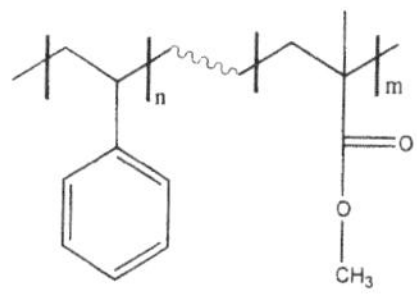	Cylinder	CdSe nanoparticles	[236]
Poly(styrene-*b*-methylmethacrylate)	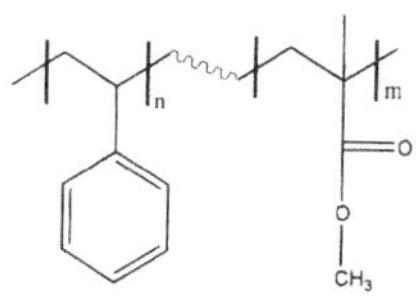	Cylinder	Nanoporous metal films	[233]
Poly(styrene-*b*-methylmethacrylate)	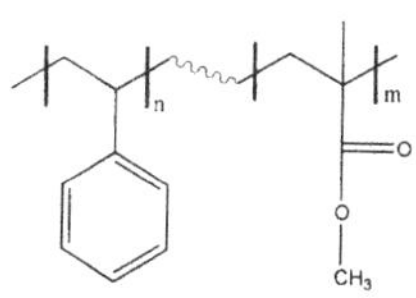	Cylinder	Nanoelectrode arrays	[241]
Poly(styrene-*b*-methylmethacrylate)	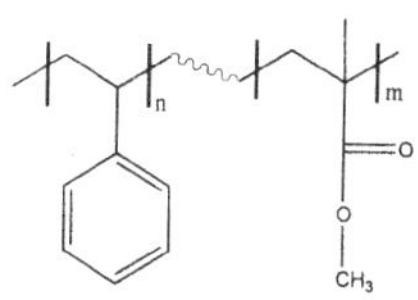	Cylinder	Surface patterned PDMS	[239]
Poly(styrene-*b*-isoprene)	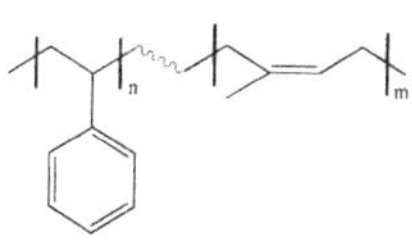	Sphere	Metal nanodots	[154, 155]

Table 5 (continued)

Block copolymer	Structure	Morphology	Application	Ref.
Poly(styrene-*b*-2-vinylpyridine)	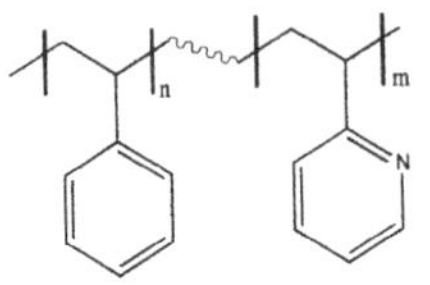	Spherical micelles	Nanoporous gold films	[159]
Poly(2,5-dioctyloxy-*para*-phenylenevinylene-*b*-styrene)	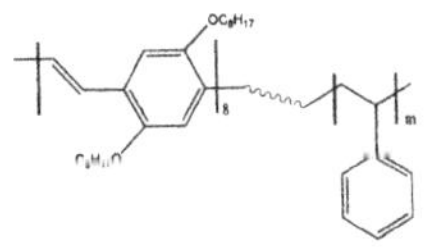	Spherical micelle	Hexagonal packed aluminum cup arrays	[35]

of nanoscale features with the formation of silicon dioxide posts [235], metal nanodots [233], CdSe nanoparticles [236], nanoporous metal films [233], nanoelectrode arrays [241] as well as surface-patterned PDMS elastomers [239].

The nanoreplication of functional nanostructures has also been achieved through other block copolymer-templated structures. De Boer et al. [35] applied honeycomb-structured films of rod–coil block copolymer as patterned templates to replicate hexagonally packed arrays of aluminum cups on the substrate surfaces (Fig. 10b). Nguyen et al. [237] embedded semiconducting polymers in the channels of oriented hexagonal nanoporous silica and used this nanoscale architecture to control the energy transfer for potential optoelectronic applications.

6 Summary and Outlook

The use of block copolymers to form a variety of different nanosized periodic patterns continues to be an active area of research. Whether in bulk, thin film, or solution micelle states, block copolymers present seemingly unlimited opportunities for fabricating and patterning nanostructures. The wealth of microstructures and the tunability of structural dimensions make them a favorable choice for scientists in a variety of research fields. As reviewed here, they can function as nanodevices themselves, or act as templates or scaffolds for the fabrication of functional nanopatterns composed of almost all types of materials. However, there are still two obvious areas which require more work: control of the long-range 3D nanostructure via more user-friendly processes and the identification of new materials with different functional properties.

Many methods to control 3D block copolymer nanostructures have been reviewed here, with particular attention paid to achieving long-range order and orientation in block copolymer thin films. However, there are still puzzles which need to be solved. For example, electric field, neutral surface, as well as eutectic solidification are effective methods to achieve large-area cylinder orientations, but the processes themselves are not trivial and the patterns obtained are still imperfect. In addition, the experimental achievement of an asymmetrical cylinder orientation is difficult compared with the symmetrical lamellar case, as exemplified by the lack of success achieved using the chemically patterned substrate approach. The combination of biaxial forces or the combination of several guiding forces has been discussed and turns out to be very effective and promising [76, 83, 134, 154]. Needless to say, a more complete understanding of the potential for interactions between different controlling methods are needed for block copolymer systems to fulfill their technical potential for nanofabrication.

In order to achieve improved nanofabrication performance, novel functional block copolymer systems are strongly desired. Many researchers have recognized this, and novel functional systems such as metal-containing block copolymer systems have significantly simplified and improved nanofabrication processes. The combination of top-down microscale patterns with the bottom-up nanopatterns are attractive for integrating functional nanostructures into multipurpose on-chip devices. However, in order to use these materials in real-time applications, further development is still needed. More ground-shaking discoveries are needed and are also fully expected.

Acknowledgements The research reported here was supported by the Nanobiotechnogy Center (NBTC), an STC Program of the National Science Foundation under Agreement No. ECS-9876771. We also thanks the Cornell Center for Materials Research (CCMR) and Cornell Nanofabrication Center (CNF) for use of their facilities.

References

1. Whitesides GM, Mathias JP, Seto CT (1991) Science 254:1312
2. Tolles WM (1996) Nanotechnology 7:59
3. Rohrer H (1996) Microelectron Eng 32:5
4. Xia Y, Rogers JA, Paul KE, Whitesides GM (1999) Chem Rev 99:1823
5. Mansky P, Harrison CK, Chaikin PM, Register RA, Yao N (1996) Appl Phys Lett 68:2586
6. Melosh NA, Davidson P, Chmelka BF (2000) J Am Chem Soc 122:823
7. Bates FS, Fredrickson GH (1990) Ann Rev Phys Chem 41:525
8. Hamley IW (1998) The physics of block copolymers. Oxford Science, p 26
9. Leibler L (1980) Macromolecules 13:1602
10. Lee M, Cho B-K, Zin W-C (2001) Chem Rev 101:3869
11. Klok H-A, Lecommandoux S (2001) Adv Mater 13:1217
12. Ober CK, Gabor AH, Gallagher-Wetmore P, Allen RD (1997) Adv Mater 9:1039
13. Fredrickson GH, Bates FS (1996) Annu Rev Mater Sci 26:501
14. Fasolka MJ, Mayes AM (2001) Annu Rev Mater Res 31:323

15. Tuzar Z, Kratochvil P (1993) Micelles of block and graft copolymers in solutions. In: Surface and colloid science, vol 15. Plenum, New York, p 1
16. Matsen MW, Bates FS (1996) Macromolecules 29:1091
17. Bates FS, Fredrickson GH (1999) Block copolymers-designer soft materials. Physics Today 52:32
18. Khandpur AK, Föerster S, Bates FS, Hamley IW, Ryan AJ, Bras W, Almdal K, Mortensen K (1995) Macromolecules 28:8796
19. Breiner U, Krappe U, Stadler R (1996) Macromol Rapid Commun 17:567
20. Breiner U, Krappe U, Abetz V, Stadler R (1997) Macromol Chem Phys 198:1051
21. Krappe U, Stadler R, Voight-Martin I (1995) Macromolecules 28:4558
22. Stadler R, Auschra C, Beckmann J, Krappe U, Voight-Martin I, Leibler L (1995) Macromolecules 28:3080
23. Matsen MW (1998) J Chem Phys 108:785
24. Abetz V (2003) Block copolymers, ternary triblocks. In: Kroschwitz JI (ed) Encyclopedia of polymer science and technology, vol 1, 3rd edn. Wiley, New York
25. Milner ST (1994) Macromolecules 27:2333
26. Zielinski JM, Spontak RJ (1992) Macromolecules 25:653
27. Radzilowski LH, Wu JL, Stupp SI (1993) Macromolecules 26:879
28. Chen JT, Thomas EL, Ober CK, Mao G-P (1996) Science 273:343
29. Cohen RE, Cheng P-L, Douzinas K, Kofinas P, Berney CV (1990) Macromolecules 23:324
30. Hamley IW, Fairclough JPA, Bates FS, Ryan AJ (1997) Polymer 39:1429
31. Muthukumar M, Ober CK, Thomas EL (1997) Science 277:1225
32. Widawski G, Rawiso M, Francois B (1994) Nature 369:387
33. Jenekhe SA, Chen XL (1999) Science 283:372
34. Hayakawa T, Horiuchi S (2003) Angew Chem Int Ed 42:2285
35. de Boer B, Stalmach U, Nijland H, Hadziioannou G (2000) Adv Mater 12:1581
36. Jenekhe SA, Chen XL (1998) Science 279:1903
37. Stupp SI, LeBonheur V, Walker K, Li LS, Huggins KE, Keser M, Amstutz A (1997) Science 276:384
38. Zubarev ER, Pralle MU, Sone ED, Stupp SI (2001) J Am Chem Soc 123:4105
39. Marsitzky D, Brand T, Geerts Y, Klapper M, Müllen K (1998) Macromol Rapid Commun 19:385
40. Gallot B (1996) Prog Polym Sci 21:1035
41. Fasolka MJ, Banerjee P, Mayes AM, Pickett G, Balazs AC (2000) Macromolecu les 33:5702
42. Mansky P, Russell TP, Hawker CJ, Pitsikalis M, Mays J (1997) Macromolecules 30:6810
43. Thurn-Albrecht T, Schotter J, Kästle GA, Emley N, Shibauchi T, Krusin-Elbaum L, Guarini K, Black CT, Tuominen MT, Russell TP (2000) Science 290:2126
44. Park M, Harrison C, Chaikin PM, Register RA, Adamson DH (1997) Science 276:1401
45. Radzilowski LH, Carvalho BL, Thomas EL (1996) J Polym Sci, Part B: Polym Phys 34:3081
46. Huinink HP, Brokken-Zijp JCM, van Dijk MA, Sevink GJA (2000) J Chem Phys 112:2452
47. Chu B, Zhou Z (1996) Physical chemistry of polyoxyalkylene block copolymer surfactant. In: Nonionic surfactants, polyoxyalkylene block copolymers, vol 60. Marcel Dekker, New York
48. Liu T, Burger C, Chu B (2003) Prog Polym Sci 28:5
49. Spatz JP, Sheiko S, Möller M (1996) Adv Mater 8:513

50. Selb J, Gallot Y (1985) Ionic block copolymers. In: Developments in block copolymers, vol 2. Elsevier, London, p 27
51. van Dijk MA, van den Berg R (1995) Macromolecules 28:6773
52. Kim H-C, Russell TP (2001) J Polym Sci Part B, Polym Phys 39:663
53. Vigild ME, Almdal K, Mortensen K, Hamley IW, Fairclough JPA, Ryan AJ (1998) Macromolecules 31:5702
54. Chen Z-R, Kornfield JA, Smith SD, Grothaus JT, Stakowski MM (1997) Science 277:1248
55. Laurer JH, Pinheiro BS, Polis DL, Winey KI (1999) Macromolecules 32:4999
56. Keller A, Pedemonte E, Willmouth FM (1970) Nature 225:538
57. Lee HH, Register RH, Hajduk DA, Bruner SM (1997) Polym Eng Sci 36:1414
58. Quiram DJ, Register RA, Marchand GR, Adamson DH (1998) Macromolecules 31:4891
59. Albalak RJ, Thomas EL (1993) J Polym Sci, Part B Polym Phys 31:37
60. Bodycomb J, Funaki Y, Kimishima K, Hashimoto T (1999) Macromolecules 32:2075
61. Amundson K, Helfand E, Davis DD, Quan X, Patel SS, Smith SD (1991) Macromolecules 24:6546
62. Amundson K, Helfand E, Quan X, Smith SD (1993) Macromolecules 26:2698
63. Amundson K, Helfand E, Quan X, Hudson SD, Smith SD (1994) Macromolecules 27:6559
64. Böker A, Knoll A, Elbs H, Abetz V, Müller AHE, Krausch G (2002) Macromolecules 35:1319
65. Chao C-Y, Li X, Ober CK, Osuji C, Thomas EL (2004) Adv Fun Mater 14:364
66. Fasolka MJ, Harrison DJ, Mayes AM, Yoon M, Mochrie SGJ (1997) Phys Rev Lett 79:3018
67. Kellogg GJ, Walton DG, Mayes AM, Lambooy P, Russell TP, Gallagher PD, Satija SK (1996) Phys Rev Lett 76:2503
68. Huang E, Russell TP, Harrison C, Chaikin PM, Register RA, Hawker CJ, Mays J (1998) Macromolecules 31:7641
69. Huang E, Rockford L, Russell TP, Hawker CJ (1998) Nature 395:757
70. Harrison C, Chaikin PM, Huse DA, Register RA, Adamson DH, Daniel A, Huang E, Mansky P, Russell TP, Hawker CJ, Egolf DA, Melnikov IV, Bodenschatz, E (2000) Macromolecules 33:857
71. Mansky P, Liu Y, Huang E, Russell TP, Hawker C (1997) Science 275:1458
72. Peters RD, Yang XM, Kim TK, Sohn BH, Nealey PF (2000) Langmuir 16:4625
73. Rockford L, Liu Y, Mansky P, Russell TP, Yoon M, Mochrie SGJ (1999) Phys Rev Lett 82:2602
74. Rockford L, Mochrie SGJ, Russell TP (2001) Macromolecules 34:1487
75. Yang XM, Peters RD, Nealey PF, Solak HH, Cerrina F (2000) Macromolecules 33:9575
76. Kim SO, Solak HH, Stoykovich MP, Ferrier NJ, de Pablo JJ, Nealey PF (2003) Nature 424:411
77. Morkved TL, Lu M, Urbas AM, Ehrichs EE, Jaeger HM, Mansky P, Russell TP (1996) Science 273:931
78. Thurn-Albrecht T, Steiner R, DeRouchey J, Stafford CM, Huang E, Bal M, Tuominen M, Hawker CJ, Russell TP (2000) Adv Mater 12:787
79. Thurn-Albrecht T, DeRouchey J, Russell TP, Jaeger HM (2000) Macromolecules 33:3250
80. Mansky P, DeRouchey J, Russell TP, Mays J, Pitsikalis M, Morkved T, Jaeger H (1998) Macromolecules 31:4399
81. Xu T, Hawker CJ, Russell TP (2003) Macromolecules 36:6178

82. Xu T, Goldbach JT, Russell TP (2003) Macromolecules 36:7296
83. Asakawa K, Hiraoka T, Hieda H, Sakurai M, Kamata Y, Naito D (2002) J Photopolym Sci Technol 15:465
84. Park C, De Rosa C, Thomas EL (2001) Macromolecules 34:2602
85. Segalman RA, Yokoyama H, Kramer EJ (2001) Adv Mater 13:1152
86. Segalman RA, Hexemer A, Kramer EJ (2003) Macromolecules 36:6831
87. Hahn J, Webber SE (2004) Langmuir 20:1489
88. Heier J, Kramer EJ, Walheim S, Krausch G (1997) Macromolecules 30:6610
89. Heier J, Genzer J, Kramer EJ, Bates FS, Walheim S, Krausch G (1999) J Chem Phys 111:11101
90. Cheng JY, Ross CA, Thomas EL, Smith HI, Vancso GJ (2002) Appl Phys Lett 81:3657
91. Cheng JY, Ross CA, Thomas EL, Smith HI, Vancso GJ (2003) Adv Mater 15:1599
92. Massey JA, Winnik MA, Manner I, Chan VZ-H, Ostermann JM, Enchelmaier R, Spatz JP, Möller M (2001) J Am Chem Soc 123:3147
93. Temple K, Kulbaba K, Power-Billard KN, Manners I, Leach KA, Xu T, Russell TP, Hawker CJ (2003) Adv Mater 15:297
94. Elhadj S, Woody JW, Niu VS, Saraf RF (2003) Appl Phys Lett 82:871
95. De Rosa C, Park C, Thomas EL, Lotz B (2000) Nature 405:433
96. Reiter G, Castelein G, Hoerner P, Riess G, Blumen A, Sommer J-U (1999) Phys Rev Lett 83:3844
97. Hong S, MacKnight WJ, Russell TP, Gido SP (2001) Macromolecules 34:2398
98. Hong S, MacKnight WJ, Russell TP, Gido SP (2001) Macromolecules 34:2876
99. Mansky P, Chaikin P, Thomas EL (1995) J Mater Sci 30:1987
100. Turturro A, Gattiglia E, Vacca P, Viola GT (1995) Polymer 36:3987
101. Kimura M, Misner MJ, Xu T, Kim SH, Russell TP (2003) Langmuir 19:9910
102. Park C, Cheng JY, Fasolka MJ, Mayers AM, Ross CA, Thomas EL, De Rosa C (2001) Appl Phys Lett 79:848
103. Lin ZQ, Kim DH, Wu XD, Boosahda L, Stone D, LaRose L, Russell TP (2002) Adv Mater 14:1373
104. Fukunaga K, Elbs H, Magerle R, Krausch G (2000) Macromolecules 33:947
105. Kim G, Libera M (1998) Macromolecules 31:2569
106. Ober CK, Li M, Douki K, Goto K, Li X (2003) J Photopolym Sci Technol 16:347
107. Du P, Li M, Douki K, Li X, Garcia CBW, Jain A, Smilgies DM, Fetters LJ, Gruner SM, Wiesner U, Ober CK (2004) Adv Mater 16:953
108. Li M, Douki K, Goto K, Li X, Coenjarts C, Smilgies DM, Ober CK (2004) Chem Mater 16:3800
109. Walton DG, Kellogg GJ, Mayes AM, Lambooy P, Russell TP (1994) Macromolecules 27:6225
110. Matsen MW (1997) J Chem Phys 106:7781
111. Geisinger T, Muller M, Binder K (1999) J Chem Phys 111:5251
112. Koneripalli N, Levicky R, Bates FS, Ankner J, Kaiser H, Satija SK (1996) Langmuir 12:6681
113. Pickett GT, Balazs AC (1997) Macromolecules 30:3097
114. Russell TP, Menelle A, Anastasiadis SH, Satija SK, Majkrzak CF (1991) Macromolecules 24:6263
115. Russell TP, Coulon G, Deline VR, Miller DC (1989) Macromolecules 22:4600
116. Fredrickson GH (1987) Macromolecules 20:2535
117. Koneripalli N, Singh N, Levicky R, Bates FS, Gallagher PD, Satija SK (1995) Macromolecules 28:2897
118. Anastasiadis SH, Russell TP, Satija SK, Majkrzak CF (1989) Phys Rev Lett 62:1852

119. Henkee CS, Thomas EL, Fetters LJ (1988) J Mater Sci 23:1685
120. Morkved TL, Jaeger HM (1997) Europhys Lett 40:643
121. Pickett GT, Witten TA, Nagel SR (1993) Macromolecules 26:3194
122. Genzer J, Kramer EJ (1997) Phys Rev Lett 78:4946
123. Thurn-Albrecht T, DeRouchey J, Russell TP, Kolb R (2002) Macromolecules 35:8106
124. Ober CK, Gabor AH (1996) J Photopoly Sci Tech 9:1
125. Reichmanis E, Ober CK, MacDonald SA, Iwayanagi T, Nishikudo T (eds) (1995) Microelectronics technology: Polymers in advanced imaging and packaging. ACS Symposium Series 614. American Chemical Society, Washington DC
126. Sundararajan N, Yang S, Ogino K, Valiyaveettil S, Wang J, Zhou X, Ober CK, Obendorf SK, Allen RD (2000) Chem Mater 12:41
127. Hartney MA, Novembre AE, Bates FS (1985) J Vac Sci Technol B 3:1346
128. DeSimone JM, York GA, McGrath JE, Gozdz AS, Bowden MJ (1991) Macromolecules 24:5330
129. Gabor AH, Lehner EA, Mao G, Schneggenburger LA, Ober CK (1994) Chem Mater 6:927
130. Vasilev C, Heinzelmann H, Reiter G (2004) J Polym Sci B: Polym Phys 42:1312
131. Trimbach D, Feldman K, Spencer ND, Broer DJ, Bastiaansen CWM (2003) Langmuir 19:10957
132. Podariu I, Chakrabarti A (2000) J Chem Phys 113:6423
133. Wang Q, Nath SK, Graham MD, Nealey PF, de Pablo JJ (2000) J Chem Phys 112:9996
134. Deng T, Ha Y-H, Cheng JY, Ross CA, Thomas EL (2002) Langmuir 18:6719
135. Yang P, Rizvi AH, Messer B, Chmelka BF, Whitesides GM, Stucky GD (2001) Adv Mater 13:427
136. Yang P, Wirnsberger G, Huang HC, Cordero SR, MaGehee MD, Scott B, Deng T, Whitesides GM, Chmelka BF, Buratto SK, Stucky GD (2000) Science 287:465
137. Fang H, Lu Y, Stump A, Reed ST, Baer T, Schunk R, Perez-Luna V, López GP, Brinker CJ (2000) Nature 405:56
138. Pereira GG, Williams DRM (1998) Macromolecules 31:5904
139. Pereira GG, Williams DRM (1999) Macromolecules 32:758
140. Wang Q, Nealey PF, de Pablo JJ (2003) Macromolecules 36:1731
141. Pereira GG, Williams DRM (1998) Phys Rev Lett 80:2849
142. Petera D, Muthukumar M (1998) J Chem Phys 109:5101
143. Huang KL, Balazs AC (1991) Phys Rev Lett 66:620
144. Balazs AC, Huang K, McElwain P, Brady JE (1991) Macromolecules 24:714
145. Chakrabarti A, Chen H (1998) J Polym Sci Part B: Polym Phys 36:3127
146. Peters RD, Yang XM, Wang Q, de Pablo JJ, Nealey PF (2000) J Vac Sci Technol B 18:3530
147. Bal M, Ursache A, Tuominen MT, Goldbach JT, Russell TP (2002) Appl Phys Lett 81:3479
148. Spatz JP, Chan VZ-H, Mößmer S, Kamm F-M, Plettl A, Ziemann P, Möller M (2002) Adv Mater 14:1827
149. Glass R, Arnold M, Blümmel J, Küller A, Möller M, Spatz JP (2003) Adv Funct Mater 13:569
150. Murray CB, Kagan CR, Bawendi MG (1995) Science 270:1335
151. Turner SW, Perez AM, Lopez A, Craighead HG (1998) J Vac Sci Technol B 16:3835
152. Harrison C, Park M, Chaikin PM, Register RA, Adamson DH (1998) J Vac Sci Technol B 16:544
153. Cheng JY, Ross CA, Chan VZ-H, Thomas EL, Lammertink RGH, Vancso GJ (2001) Adv Mater 13:1174

154. Park M, Chaikin PM, Register RA, Adamson DA (2001) Appl Phys Lett 79:257
155. Li RR, Dapkus PD, Thompson ME, Jeong WG, Harrison C, Chaikin PM, Register RA, Adamson DH (2000) Appl Phys Lett 76:1689
156. Black CT, Guarini KW, Milkove KR, Baker SM, Russell TP, Tuominen MT (2001) Appl Phys Lett 79:409
157. Guarini KW, Black CT, Milkove KR, Sandstrom RL (2001) J Vac Sci Technol B 19:2784
158. Spatz JP, Herzog T, Mößmer S, Ziemann P, Möller M (1999) Adv Mater 11:149
159. Haupt M, Miller S, Glass R, Arnold M, Sauer R, Thonke K, Möller M, Spatz JP (2003) Adv Mater 15:829
160. Lammertink RGH, Hempenius MA, van den Enk JE, Chan VZ-H, Thomas EL, Vancso GJ (2000) Adv Mater 12:98
161. Russell TP, Thurn-Albrecht T, Tuominen M, Huang E, Hawker CJ (2000) Macromol Symp 159:77
162. Lammertink RGH, Hempenius MA, Thomas EL, Vancso GJ (1999) J Polym Sci, Polym Phys Ed 37:1009
163. Lammertink RGH, Hempenius MA, Chan VZ-H, Thomas EL, Vancso GJ (2001) Chem Mater 13:429
164. Spatz JP, Eibeck P, Mößmer S, Möller M, Herzog T, Ziemann P (1998) Adv Mater 10:849
165. Koslowski B, Stobel S, Herzog T, Heinz B, Boyen HG, Notz R, Ziemann P, Spatz JP, Möller M (2000) J Appl Phys 87:7533
166. Urbas A, Fink Y, Thomas EL (1999) Macromolecules 32:4748
167. Ruokolainen J, Mäkinen R, Torkkeli M, Mäkelä T, ten Brinke G, Ikkala O (1998) Science 280:557
168. Morkved TL, Wiltzius P, Jaeger HM, Grier DG, Witten TA (1994) Appl Phys Lett 64:422
169. Lopes WA, Jaeger HM (2001) Nature 414:735
170. Tadd EH, Bradley J, Tannenbaum R (2002) Langmuir 18:2378
171. Tadd EH, Bradley J, Goldberg EP, Tannenbaum R (2002) Mat Res Soc Symp Proc 707:205
172. Brown GD, Watkins JJ (2000) Mater Res Soc Symp Proc 584:169
173. Abes JI, Cohen RE, Ross CA (2003) Chem Mater 15:1125
174. Abes JI, Cohen RE, Ross CA (2003) Mater Sci Eng C23:641
175. Möller M, Spatz JP, Roescher A (1996) Adv Mater 8:337
176. Goren M, Lennox RB (2001) Nano Lett 1:735
177. Lee T, Yao N, Aksay IA (1997) Langmuir 13:3866
178. Kane RS, Cohen RE, Silbey R (1996) Chem Mater 8:1919
179. Fogg DE, Radzilowski LH, Blanski R, Schrock RR, Thomas EL (1997) Macromolecules 30:417
180. Ansari IA, Hamley IW (2003) J Mater Chem 13:2412
181. Wong MS, Cha JN, Choi K-S, Deming TJ, Stucky GD (2002) Nano Lett 2:583
182. Sone ED, Zubarev ER, Stupp SI (2002) Angew Chem Int Ed 41:1705
183. Park C, McAlvin JE, Fraser CL, Thomas EL (2002) Chem Mater 14:1225
184. Manners I (2001) Science 294:1664
185. Wang X-S, Arsenault A, Ozin GA, Winnik MA, Manners I (2003) J Am Chem Soc 125:12686
186. Kowalewski T, Tsarevsky NV, Matyjaszewski K (2002) J Am Chem Soc 124:10632
187. Wu CG, Bein T (1994) Science 264:1757
188. Martin CR (1996) Chem Mater 8:1739
189. Foss CA, Hornyak GL, Stockert JA, Martin CR (1992) J Phy Chem 96:7497

190. Kresge CT, Leonowicz ME, Roth WJ, Vartuli JC, Beck JS (1992) Nature 359:710
191. Lee H-K, Lee H, Ko YH, Chang YJ, Oh N-K, Zin W-C, Kim K (2001) Angew Chem Int Ed 40:2669
192. Krause B, Sijbesma HJP, Münüklü P, van der Vege NFA, Wessling M (2001) Macromolecules 34:8792
193. Chan VZ-H, Hoffman J, Lee VY, Iatrou H, Avgeropoulos A, Hadjichristidis N, Miller RD, Thomas EL (1999) Science 286:1716
194. Corma A (1997) Chem Rev 97:2373
195. Gin DL, Gu W (2001) Adv Mater 13:1407
196. Feng X, Fryxell GE, Wang L-Q, Kim AY, Liu J, Kemner KM (1997) Science 276:923
197. Mercier L, Pinnavaia TJ (1997) Adv Mater 9:500
198. Lee SB, Mitchell DT, Trofin L, Nevanen TK, Södelund H, Martin CR (2002) Science 296:2198
199. Yang S, Mirau PA, Pai C-S, Nalamasu O, Reichmanis E, Pai JC, Obeng YS, Seputro J, Lin EK, Lee H-J, Sun J, Gidley DW (2002) Chem Mater 14:369
200. Hedrick JL, Miller RD, Hawker CJ, Carter KR, Volksen W, Yoon DY, Trollsås M (1998) Adv Mater 10:1049
201. Huo Q, Magolese DI, Ciesla U, Feng P, Gier TE, Sieger P, Leon R, Petroff PM, Schüth F, Stucky GD (1994) Nature 368:317
202. Harrison DJ, Fluri K, Seiler K, Fan Z, Effenhauser CS, Manz A (1993) Science 261:895
203. Zhao D, Feng J, Huo Q, Melosh N, Fredrickson GH, Chmelka BF, Stucky GD (1998) Science 279:548
204. Kruk M, Jaroniec M, Ko CH, Ryoo R (2000) Chem Mater 12:1961
205. Van Grieken R, Calleja G, Stucky GD, Melero JA, Garcia RA, Iglesias J (2003) Langmuir 19:3966
206. Yang C-M, Zibrowius B, Schmidt W, Schüth F (2003) Chem Mater 15:3739
207. Kriesel JW, Sander MS, Tilley TD (2001) Chem Mater 13:3554
208. Yang P, Zhao D, Margolese DI, Chemlka BF, Stucky GD (1998) Nature 396:152
209. Templin M, Franck A, Chesne AD, Leist H, Zhang Y, Ulrich R, Schädler V, Wiesner U (1997) Science 278:1795
210. Krämer E, Förster S, Göltner C, Antonietti M (1998) Langmuir 14:2027
211. Lee J-S, Hirao A, Nakahama S (1988) Macromolecules 21:274
212. Xu T, Stevens J, Villa JA, Goldbach JT, Guarini KW, Black CT, Hawker CJ, Russell TP (2003) Adv Fun Mater 13:698
213. Jeong U, Ryu DY, Kim JK, Kim DH, Russell TP, Hawker CJ (2003) Adv Mater 15:1247
214. Jeong U, Kim H-C, Rodriguez RL, Tsai IY, Stafford CM, Kim JK, Hawker CJ, Thomas TP (2002) Adv Mater 14:274
215. Liu G, Ding J, Hashimoto T, Kimishima K, Winnik FM, Nigam S (1999) Chem Mater 11:2233
216. Zalusky AS, Olayo-Valles R, Wolf JH, Hillmyer MA (2002) J Am Chem Soc 124:12761
217. Zalusky AS, Olayo-Valles R, Taylor CJ, Hillmyer MA (2001) J Am Chem Soc 123:1519
218. Wolf JH, Hillmyer MA (2003) Langmuir 19:6553
219. Mäki-Ontto R, de Moel K, de Odorico W, Ruokolainen J, Stamm M, ten Brinke G, Ikkala O (2001) Adv Mater 13:117
220. Antonietti M, Göltner C (1997) Angew Chem Int Ed 36:910
221. Ying JY, Mehnert CP, Wong MS (1999) Angew Chem Int Ed 38:56
222. Förster S, Antonietti M (1998) Adv Mater 10:195
223. Golner CG, Antonietti M (1997) Adv Mater 9:431
224. Thomas A, Schlaad H, Smarsly B, Antonietti M (2003) Langmuir 19:4455
225. Zhao D, Yang P, Melosh N, Feng J, Chmelka BF, Stucky GD (1998) Adv Mater 10:1380

226. Pal RA, Humayun R, Schulberg MT, Sengupta A, Sun J-N, Watkins JJ (2004) Science 303:507
227. Liu G, Ding J, Stewart S (1999) Angew Chem Int Ed 38:835
228. Buchmeiser MR (2001) Angew Chem Int Ed 40:3795
229. Drzal PL, Halasa AF, Kofinas P (2000) Polymer 41:4671
230. Pitois O, Francois B (1999) Colloid Polym Sci 277:574
231. Karthaus O, Maruyama N, Cieren X, Shimomura M, Hasegawa H, Hashimoto T (2000) Langmuir 16:6071
232. Kesting RE (1985) Synthetic polymer membranes. Wiley, New York
233. Shin K, Leach KA, Goldbach JT, Kim DH, Jho JY, Tuominen M, Hawker CJ, Russell TP (2002) Nano Lett 2:933
234. Parthasarathy RV, Phani KLN, Martin CR (1995) Adv Mater 7:896
235. Kim H-C, Jia X, Stafford CM, Kim DH, McCarthy TJ, Tuominen M, Hawker CJ, Russell TP (2001) Adv Mater 13:795
236. Misner MJ, Skaff H, Emrick T, Russell TP (2003) Adv Mater 15:221
237. Nguyen T-Q, Wu J, Doan V, Schwartz BJ, Tolbert SH (2000) Science 288:652
238. Parthasarathy RV, Martin CR (1994) Nature 369:298
239. Kim DH, Lin Z, Kim H-C, Jeong U, Russell TP (2003) Adv Mater 15:811
240. Martin CR (1994) Science 266:1961
241. Jeoung E, Galow TH, Schotter J, Bal M, Ursache A, Tuominen MT, Stafford CM, Russell TP, Rotello VM (2001) Langmuir 17:6396

Author Index Volumes 101–190

Author Index Volumes 1–100 see Volume 100

Subject Index

The manufacturer's authorised representative in the EU is Springer Nature Customer Service Centre GmbH, Europaplatz 3, 69115 Heidelberg, Germany. If you have any concerns regarding our products, please contact ProductSafety@springernature.com

Printed and bound by CPI Group (UK) Ltd, Croydon, CR0 4YY
22/09/2026
02221064-0012